Biochemical and Environmental Toxicology

Biochemical and Environmental Toxicology

Poonam Pandey

AMIGA PRESS. INC
Delhi -110051

AMIGA PRESS INC.
Registered Office:
19 DBS Building, Ansari Road, Darya Ganj, Delhi. 110002, India
Phone: +91-11-23283333 Email: amigapress@gmail.com

This edition is published by Amiga Press.Inc

ISBN : 978-93-84533-13-7

First Published 2018
Reprint 2026

Printed and bound in India by Replika Press Pvt. Ltd.

Preface

Biochemistry governs all living organisms and living processes. By controlling information flow through biochemical signalling and the flow of chemical energy through metabolism, biochemical processes give rise to the incredible complexity of life. As our understanding of how various agents can cause harm to humans and other organisms, a more descriptive definition of toxicology is "the study of the adverse effects of chemicals or physical agents on living organisms".

Biological functions attributed to laccases include spore resistance and pigmentation, lignification of plant cell walls, lignin biodegradation, humus turnover and detoxification processes, virulence factors, and copper and iron homeostasis.

Laccases exhibit an extraordinary natural substrate range (phenols, polyphenols, anilines, aryl diamines, methoxy-substituted phenols, hydroxyindols, benzenethiols, inorganic/organic metal compounds and many others) which is the major reason for their attractiveness for dozens of biotechnological applications. Moreover, in the presence of small molecules, known as redox mediators, laccases enhance their substrate specificity. Indeed, laccase oxidizes the mediator and the generated radical oxidizes the substrate by mechanisms different from the enzymatic one, enabling the oxidative transformation of substrates with high redox potentials-otherwise not oxidized by the enzyme. The industrial applicability of laccase may therefore be extended by the use of a laccase-mediator system (LMS). Thus, laccase and LMS find potential application in delignification and biobleaching of pulp; treatment of wastewater from industrial plants; enzymatic modification of fibres and dye-bleaching in the textile and dye industries; enzymatic crosslinking of lignin-based materials to produce medium density fibreboard; detoxification of pollutants and bioremediation; detoxification of lignocellulose hydrolysates for ethanol production by yeast; enzymatic removal of phenolic compounds in beverages-wine and beer stabilization, fruit juice processing-; and construction of biosensors and biofuel cells.

A major area of research in the field of environmental toxicology involves how toxic elements released into the environment precipitate the emergence of cancers such as leukemia, lung cancer, lymphomas, and melanoma. An environmental toxicologist will investigate the incidences of such cancers and how intensely exposed an individual or groups of people suffering from cancer have been to certain chemicals. Some of these harmful chemicals include pesticides, asbestos, solvents, and herbicides. In addition, environmental toxicologists are often activists who try to lobby for regulations on these chemicals.

Environmental remediation involves providing a remedy for an environmental problem. This can include removing contaminants from groundwater or cleaning up after an oil spill. Remediation is not always just subject to the will of the people, but is often a matter of government regulation or intervention.

One of the main purposes of environmental remediation is to restore contaminated sites or resources to a level that is safe for humans and animals. Depending on the type of damage that is done, this can be a complex and expensive process. There are companies that specialize in environmental remediation. Even with the help of these professionals and environmental experts, however, sometimes there is nothing that can be done to restore a contaminated site to a point where it is safe.

Toxicology is an evolving medical science and so is toxicology terminology. The book will be of immense value to the students of environment, botany, zoology, biology, agriculture, biochemistry and medicine.

—Editor

Contents

1

The Fundamental Principle of Toxicology

Early humans recognised basically two types of substances—beneficial ones (water, food, and medicine) and harmful ones (poisons). However, more recent science has shown that virtually every substance is toxic, although the quantity or dose needed to cause specific effects varies considerably. Paracelsus stated this idea nearly five centuries ago: "All substances are poisons; there is none that is not a poison. The right dose differentiates a poison and a remedy." For example, two tablets of Extra-Strength Tylenol® (acetaminophen) may relieve a headache, but 40 tablets (20 grams) taken at once can cause serious liver injury or death. Even plain water can be a fatal poison. One German toxicologist reported on a man who died from cerebral edema and severe electrolyte imbalance after drinking 17 litres of water within a very short time.

Chronic vs. Acute Exposure

Health effects caused by exposure to toxic substances are usually differentiated based on whether the adverse effect occurs after long-term (chronic) exposure or short-term (acute) exposure. Some substances, such as cyanide, are acutely toxic, which means they cause rapid death after a brief exposure to a lethal (often small) dose. The effects of other substances, such as crocidolite asbestos, are more often cumulative, causing measurable damage only after years of exposure.

How Toxins Work (The Receptor Theory)

All living organisms operate through highly integrated sets of biochemical reactions, which are sensitive to conditions including temperature, pH, and the concentration of other chemicals in the system. Some chemicals, such as strong acids and bases, are toxic simply because they denature proteins and dissolve living tissue. Other chemicals, however, exert their toxic effect by binding with specific receptors in cells, thereby disrupting normal biochemical reactions.

For example, carbon monoxide interferes with human respiration because it has a much stronger affinity for hemoglobin than oxygen does. With the hemoglobin's oxygen receptors tied up, the body cannot process sufficient oxygen, and death may occur. In multicellular animals, including humans, the body has many finely tuned regulatory systems to ensure things work properly in response to external conditions. This tendency of an organism to maintain normal stability or equilibrium is called homeostasis. Some things can disrupt homeostasis, such as external physical circumstances (extreme heat or cold) or internal chemical or biological agents. For example, organophosphate compounds (including the deadly nerve gas sarin and a common insecticide called parathion) affect homeostasis by deactivating an enzyme responsible for the uptake of the key neurotransmitter acetylcholine. The nervous system becomes overstimulated and induces the sweat glands to produce sweat and the body's muscles to twitch or convulse. It is this loss of homeostasis that produces the symptoms of poisoning or disease.

PRINCIPLES OF TOXICOLOGY

Instructional Objectives

The student information regarding the classification of poisons, types of poisoning, metabolism of toxicants, mode of action of toxicants and factors affecting the action of toxicants. In addition, common causes of toxicosis as well as common toxicological problems in domestic and farm animals.

Performance Objectives

Students should have a broad understanding of toxicants especially with regard to their physiological effect on the body. The students should also be well acquainted with commonly encountered toxicoses in selected domestic and farm animals including various symptoms associated with these toxicoses and specific lesions associated with various types of toxicoses.

Classification of Poisons

The graduate veterinarian's ability to serve his clientele is no better than his ability to accurately differentiate between the various possible causes of the illnesses presented to him. It should be noted that mere memorization of clinical signs is not the only way to recall a certain toxicological problem. Perhaps, a better method of recognition is to classify the common toxicoses according to their specific lesions. Since there are many plants, feed additives, insecticides, herbicides, fungicides, and heavy metals that may fall under a specific lesion, each lesion could be subdivided and indexed to help with differential diagnosis. The following classification based on toxic effects of poisons is far from complete, but it may act as a starting point for future indices of diagnostic classifications.

Gastrointestinal Syndrome

This is seen in poisoning from oak, acute copper, E.coli, Salmonella, carbamate fungicides, ANTU, acute crotalaria, paraquat herbicides, thallium, bracken fern, warfarin, etc. The important clinical signs include: bloody diarrhea, hematuria, enteritis, sometimes GI irritation. Death may occur within one to a few days. In case of wild indigo, staphylococcus, peptides and amides, and borates deaths seldom occur. Severe gastrointestinal syndrome is caused by pokeweed, sneezweed, inorganic arsenic herbicides. Death may occur within 1-2 days with bloody diarrhea, hemolytic crisis (pokeweed), and excess salivation (sneezweed).

Neuromuscular (Paresis) Effects

Some poisons causing these effects act rapidly, some act slowly. Oleander, buttercup, buckeye, lead(in poultry)act rapidly causing death in 1-2 days. Others include: death comas, poison hemlock, organophosphates. All the above cause in coordination plus GI syndrome where as jimsonweed,2,4-D, organic tin, ticks cause only in coordination. Other examples of toxicants causing CNS effects include botulism, horse tail, bracken fern (in horses), pigweed, etc.

Bone-Teeth-Hoof-Hair Deformities

Ergot - sloughing of tips of tail, ears, teats, etc.

Fluoride - bone and teeth lesions

Selenium (chronic), poisonvetch, copper deficiency, chronic thallium, lead nitrate, chlorinated naphthalene, are all associated with hair problem.

Kidney Lesions

Depending on the types of renal damage caused by toxicants kidney lesions are categorized as follows:

- *Hematuria*: This may be accompanied by fever as in the case of bracken fern or there may be no fever as seen in poisoning from oak, cassia, inorganic mercury or cadmium.
- *Hemoglobinuria*: This is caused by crucifers(mustard) with photosensitization. Similar condition is seen in chronic copper toxicity.

- *Oxalates*: These are formed in poisoning from amaranthus (which also causes perirenal edema), halogeton, black greasewood, rhubarb, and ethylene glycol.
- *Perirenal Edema*: Amaranthus and oak(with as cites); nightshade (no as cites).
- *Degenerative Changes in Kidneys*: These can be caused by chronic organic mercury compounds, chronic thallium, sulfonamides and so forth.

Liver Lesions

These are caused by several toxicants such as selenium (acute,) aflatoxins, crotalaria, tannic acid, carbon tetrachloride, phenothiazines, gossypol, tarweed, etc.

Photosensitization

Phenolic fungicides (necrosis of contact tissue); vehicles for insecticides; blue-green algae(death within 1-2 days); crucifers (with hemoglobinuria), snow-on-the-mountain (without hemoglobinuria); phenothiazines, alfalfa; St. Johnswort, horse brush.

Hemorrhagic Syndrome

Bracken fern (with fever); those that do not have fevers include: sweet clover, warfarin, pindone, Radiation, mycotoxins, crotalaria.

Fevers

Castorbean, oleander, have rapidly acting toxic materials and death may occur within 1-2 days. On the other hand, milkweed, bracken fern, buckeye, locust, etc. are known to be slow-acting.

Abortions and/or Anomalies

Fusarium fungal metabolities, poisonvetch, lupine, broomweed, ergot,2,4,5-T,nitrates. The following cause birth defects: chronic selenium, veratrum, oak, locoweed, jimsonweed, hem- lock, lead, and mercury.

The classification of poisons made are based on the organ or system that is the 'target' site for the effect of the toxic chemical. Such classification that is based on the toxic effects of poisons on the body may be unsatisfactory because the same substance can have different effects on different organs of the body. It can also vary in its action between one species and another. Therefore a relatively simple classification can be made on the basis of the structural features of the toxic chemicals that are responsible for their toxic properties and their affinity for 'target' sites in the animal body. Various diverse chemical compounds can be thus divided into two groups, namely, inorganic and organic compounds.

Inorganic compounds include metals, metalloids, their salts and acids and alkalis. Organic compounds on the other hand include all carbon compounds other than carbonates, and the metallic carbides and cyanides. Furthermore, an analytical toxicologist endeavours to separate poisons into characteristic groups and several classifications can be made according to the analytical procedures involved. A typical subdivision is:

- Volatile poisons,
- Metallic poisons,
- Toxic anions,
- Non-volatile organic poisons isolated by solvent extractions,
- Miscellaneous poisons.

Finally, poisons may often be classified conveniently by their origin (plant poisons) or use (pesticides).

SUB-DISCIPLINES OF TOXICOLOGY

The field of toxicology can be further divided into the following sub-disciplines or subspecialties:

- Environmental Toxicology is concerned with the study of chemicals that contaminate food, water, soil, or the atmosphere. It also deals with toxic substances that enter bodies of waters such as lakes, streams, rivers, and oceans. This sub-discipline addresses the question of how various plants, animals, and humans are affected by exposure to toxic substances.
- Occupational (Industrial) Toxicology is concerned with health effects from exposure to chemicals in the workplace. This field grew out of a need to protect workers from toxic substances and to make their work environment safe. Occupational diseases caused by industrial chemicals account for an estimated 50,000 to 70,000 deaths, and 350,000 new cases of illness each year in the United States.
- Regulatory Toxicology gathers and evaluates existing toxicological information to establish concentration-based standards of "safe" exposure. The standard is the level of a chemical that a person can be exposed to without any harmful health effects.
- Food Toxicology is involved in delivering a safe and edible supply of food to the consumer. During processing, a number of substances may

be added to food to make it look, taste, or smell better. Fats, oils, sugars, starches and other substances may be added to change the texture and taste of food. All of these additives are studied to determine if and at what amount, they may produce adverse effects. A second area of interest includes food allergies. Almost 30 per cent of the American people have some food allergy. For example, many people have trouble digesting milk, and are lactose intolerant. In addition, toxic substances such as pesticides may be applied to a food crop in the field, while lead, arsenic, and cadmium are naturally present in soil and water, and may be absorbed by plants. Toxicologists must determine the acceptable daily intake level for those substances.

- Clinical Toxicology is concerned with diseases and illnesses associated with short term or long term exposure to toxic chemicals. Clinical toxicologists include emergency room physicians who must be familiar with the symptoms associated with exposure to a wide variety of toxic substances in order to administer the appropriate treatment.
- Descriptive Toxicology is concerned with gathering toxicological information from animal experimentation. These types of experiments are used to establish how much of a chemical would cause illness or death. The United States Environmental Protection Agency (EPA), the Occupational Safety and Health Administration (OSHA), and the Food and Drug Administration (FDA), use information from these studies to set regulatory exposure limits.
- Forensic Toxicology is used to help establish cause and effect relationships between exposure to a drug or chemical and the toxic or lethal effects that result from that exposure.
- Analytical toxicology identifies the toxicant through analysis of body fluids, stomach content, excrement, or skin.
- Mechanistic Toxicology makes observations on how toxic substances cause their effects. The effects of exposure can depend on a number of factors, including the size of the molecule, the specific tissue type or cellular components affected, whether the substance is easily dissolved in water or fatty tissues, all of which are important when trying to determine the way a toxic substance causes harm, and whether effects seen in animals can be expected in humans.

CLASSIFICATION OF TOXIC AGENTS

- *Heavy Metals*: Metals differ from other toxic substances in that they are neither created nor destroyed by humans. Their use by humans plays an important role in determining their potential for health effects.

Their effect on health could occur through at least two mechanisms: first, by increasing the presence of heavy metals in air, water, soil, and food, and second, by changing the structure of the chemical. For example, chromium III can be converted to or from chromium VI, the more toxic form of the metal.

- *Solvents and Vapors*: Nearly everyone is exposed to solvents. Occupational exposures can range from the use of "white-out" by administrative personnel, to the use of chemicals by technicians in a nail salon. When a solvent evaporates, the vapors may also pose a threat to the exposed population.
- Radiation and Radioactive Materials: Radiation is the release and propagation of energy in space or through a material medium in the form of waves, the transfer of heat or light by waves of energy, or the stream of particles from a nuclear reactor.
- *Dioxin/Furans*: Dioxin, (or TCDD) was originally discovered as a contaminant in the herbicide Agent Orange. Dioxin is also a by-product of chlorine processing in paper producing industries.
- *Pesticides*: The EPA defines pesticide as any substance or mixture of substances intended to prevent, destroy, repel, or mitigate any pest. Pesticides may also be described as any physical, chemical, or biological agent that will kill an undesirable plant or animal pest.
- *Plant Toxins*: Different portions of a plant may contain different concentrations of chemicals. Some chemicals made by plants can be lethal. For example, taxon, used in chemotherapy to kill cancer cells, is produced by a species of the yew plant.
- *Animal Toxins*: These toxins can result from venomous or poisonous animal releases. Venomous animals are usually defined as those that are capable of producing a poison in a highly developed gland or group of cells, and can deliver that toxin through biting or stinging. Poisonous animals are generally regarded as those whose tissues, either in part or in their whole, are toxic.
- *Subcategories of Toxic Substance Classifications*: All of these substances may also be further classified according to their:
  - Effect on target organs (liver, kidney, hematopoietic system).
  - Use (pesticide, solvent, food additive).
  - Source of the agent (animal and plant toxins).
  - Effects (cancer mutation, liver injury).
  - Physical state (gas, dust, liquid).
  - Labeling requirements (explosive, flammable, oxidiser).
  - Chemistry (aromatic amine, halogenated hydrocarbon).

  - Poisoning potential (extremely toxic, very toxic, slightly toxic).
- *General Classifications of Interest to Communities*:
  - Air pollutants.
  - Occupation-related.
  - Acute and chronic poisons.

All chemicals (or any chemical) may be poisonous at a given dose and through a particular route. For example, breathing too much pure oxygen, drinking excessive amounts of water or eating too much salt can cause poisoning or death.

DEALS OF TOXICOKINETICS

Toxicokinetics deals with the distribution, metabolism, and elimination of toxins in the body. Once chemicals enter the body, they are metabolised, with some eliminated and some retained. Various elimination mechanisms such as exhalation, perspiration, urination, defecation, or detoxification rid the body of the chemical over a period of time.

For some chemicals, the elimination rate is so slow that they may persist in the body for a lifetime. These retained substances may reside in fatty tissue, bones, or organs. When chemicals are absorbed, they may spread through the bloodstream in unmodified form or be transformed before reaching specific receptors.

Chemicals absorbed in the gastrointestinal tract are carried by the portal vein to the liver, where they may be altered through various metabolic reactions before being released into general circulation.

These reactions are frequently called detoxification, but the term can be misleading because the reactions can create metabolites that in some cases are more toxic than the original chemical (bioactivation).

Hydrophilic (very water-soluble) substances in the bloodstream usually do not need to be metabolised, because they can go directly to the kidneys and be excreted as urine, or they can be exhaled through the lungs if they evaporate easily.

The tear ducts and sweat glands are also methods of elimination of some substances. Substances that are more soluble in fat (lipophilic) tend to accumulate in fatty tissues. In the liver, these substances may undergo two phases of detoxification. Monooxygenation reactions introduce reactive groups that are then conjugated with sugars or other hydrophilic compounds. Depending upon other properties of the original compound, these conjugates are then excreted in the urine or in the bile, which empties into the small intestines.

TOXICOLOGICAL INFORMATION SOURCES

The Agency for Toxic Substances and Disease Registry (ATSDR)

ATSDR is part of the U.S. Department of Health and Human Services. It was created by Congress in 1980 to provide health-based information for use in the cleanup of chemical waste disposal sites mandated by the Comprehensive Environmental Response, Compensation, and Liability Act (CERCLA). As the lead agency for implementing the health-related guidelines of CERCLA, ATSDR assesses the presence and nature of health hazards at specific Superfund Sites, to help prevent or reduce further exposure and the illnesses that result from such exposures, and to expand the knowledge-base about health effects from exposure to hazardous substances.

ATSDR is mostly concerned with the health effects that may occur from exposure to toxic chemicals. ATSDR's Hazardous Substances and Health Effects Database (HazDat) discusses the issue.

ATSDR also publishes Toxicological Profiles (which provide information on specific chemicals and possible health effects), Case Studies in Environmental Medicine (which are used to provide information to health care providers about the toxic effects of chemicals), and Public Health Statements (which contain information on toxic chemical exposures). ATSDR's Division of Toxicology also produces ToxFAQsTM, a series of summaries about hazardous substances. Information for this series is excerpted from the ATSDR Toxicological Profiles and Public Health Statements.

Each ToxFAQ summary is quick and easy to understand, and answers the most frequently asked questions about exposure to hazardous substances found around hazardous waste sites and the effects of exposure on human health. Medical Management Guidelines for Acute Chemical Exposures (Guidelines) were developed by ATSDR to aid emergency department physicians and other emergency health care professionals who manage acute exposures resulting from chemical incidents.

The guidelines are intended to aid health care professionals involved in emergency response to effectively decontaminate patients, protect themselves and others from contamination, communicate with other involved personnel, efficiently transport patients to a medical facility, and provide competent medical evaluation and treatment to exposed persons.

The United States Environmental Protection Agency (EPA)

EPA is responsible for a number of activities, including enforcing federal laws designed to protect human health and the environment. There are ten regional EPA offices throughout the United States, with EPA headquarters located in Washington, D.C. Each regional office is responsible within selected states for implementing the agency's programmes, considering regional needs and implementing federal environmental laws.

Following is a list of the regions and the states they cove:

- *Region* 1: Connecticut, Maine, Massachusetts, New Hampshire, Rhode Island, Vermont.
- *Region* 2: New Jersey, New York, Puerto Rico, and the U.S. Virgin Islands.
- *Region* 3: Delaware, Maryland, Pennsylvania, Virginia, West Virginia, the District of Columbia.
- *Region* 4: Alabama, Florida, Georgia, Kentucky, Mississippi, North Carolina, South Carolina, Tennessee.
- *Region* 5: Illinois, Indiana, Michigan, Minnesota, Ohio, Wisconsin.
- *Region* 6: Arkansas, Louisiana, New Mexico, Oklahoma, Texas.
- *Region* 7: Iowa, Kansas, Missouri, Nebraska.
- *Region* 8: Colorado, Montana, North Dakota, South Dakota, Utah, Wyoming.
- *Region* 9: Arizona, California, Hawaii, Nevada, the territories of Guam and American Samoa.
- *Region* 10: Alaska, Idaho, Oregon, Washington.

The specific chemicals regulated by EPA and the standards associated with them are found in the Code of Federal Regulations or CFR. The different sections of the CFR are called Titles, and the ones that apply to EPA are in Title 40. EPA has developed rules and regulations that activate the requirements of several environmental laws.

In addition, the following EPA Laws regulate chemicals in the environment:

- The Clean Air Act implements regulations that control and abate air emissions from stationary and mobile sources.
- The Clean Water Act regulates discharge of pollutants to surface waters.
- The Safe Drinking Water Act establishes standards for contaminants in drinking water; regulates discharges to underground injection wells, sole source aquifers, and public drinking water systems.
- The Comprehensive Environmental Response, Compensation, and Liability Act (CERCLA or Superfund) deals with cleanup of hazardous waste sites and definition of equirements for response to hazardous waste spills.
- The Resource Conservation and Recovery Act (RCRA) deals with

identification and regulation of hazardous waste treatment, storage, and disposal.

- The Federal Insecticide, Fungicide, and Rodenticide Act (FIFRA) requires registration and testing of pesticides, regulates their sale, distribution, and use.
- The Toxic Substances Control Act (TSCA) requires testing and reporting of chemicals prior to manufacturing, distribution, and use; and restricts the use of chemicals that pose a threat to human health and the environment.
- The Emergency Planning and Community Right to Know Act (EPCRA) requires companies to report inventories of hazardous chemicals and toxic releases; and requires state and local governments to develop plans for responding to emergency releases.

In addition, EPA's Office of Research and Development (ORD) studies the effects of toxic exposure on people and the environment.

The Centers for Disease Control and Prevention (CDC)

C DC is an agency in the Department of Health and Human Services, its mission is to promote health and quality of life by preventing and controlling disease, injury, and disability. In the past, the CDC has focused on the study and prevention of infectious diseases such as malaria and smallpox. However, now its responsibilities have enlarged to include environmental and occupational hazards.

The CDC Centers that deal with environmental health are the National Center for Environmental Health (NCEH) and the National Institute for Occupational Safety and Health (NIOSH). NCEH addresses hazards associated with chemical exposure inside and outside the workplace. NIOSH was established by the Occupational Safety and Health Act of 1970. NIOSH has several functions, including investigating potentially hazardous work conditions, and evaluating chemical hazards in the workplace. NIOSH is the only federal institute responsible for conducting research and making recommendations for the prevention of work-related illnesses and injuries.

NIOSH's responsibilities include:

- Investigating potentially hazardous working conditions as requested by employers or employees.
- Evaluating hazards in the workplace, ranging from chemicals to machinery.
- Creating and disseminating methods for preventing disease, injury, and disability.

- Conducting research and providing scientifically valid recommendations for protecting workers.
- Providing education and training to individuals preparing for or actively working in the field of occupational safety and health.

Information gathered from these activities is used to help reduce disease, injury and disability in the workplace. The information is provided to OSHA, which uses it to establish standards to protect health in the workplace.

The Nuclear Regulatory Commission (NRC)

NRC, established in 1974, regulates the use of nuclear materials for commercial, industrial, academic, and medical purposes. This includes regulating nuclear power plants, nuclear materials used in the diagnosis and treatment of cancer, and nuclear materials used in smoke detectors. NRC also regulates non-power research, test and training reactors; nuclear fuel cycle facilities (the production of nuclear fuel); and the transport, storage, and disposal of nuclear materials and waste. Like OSHA and EPA, NRC obtains and evaluates information about acceptable exposure levels for workers handling nuclear materials.

The Food and Drug Administration (FDA)

FDA promotes and protects the public health by helping safe and effective products reach the market in a timely way, and monitoring products for continued safety after they are in use.

The American Conference of Governmental Industrial Hygienists (ACGIH)

ACGIH is a professional organisation that produces a listing of Threshold Limit Values (TLV) and Biological Exposure Indices (BEI) for several hundred chemicals, updating them every year. BEIs are recommended maximum concentrations of various types of toxic substances, and are guidelines to evaluate the potential health hazards associated with exposure. The maximum concentration may be measured in blood, urine, or exhaled air. The TLVs are guidelines for occupational exposure to hazardous chemicals and are published in a booklet called Threshold Limit Values (TLVs) for Chemical Substances and Physical Agents and Biological Exposure Indices.

Electronic Databases

Information on toxic chemicals is available at the following sites:

- Toxicology Data Network: Several databases, such as the Hazardous Substances Data Bank (HSDB) and the Registry of Toxic Effects of

Chemical Substances (RTECS), can be found in the TOXNET database. The HSDB has toxicological information on more than 4,000 chemicals, as well as information on emergency handling procedures, environmental data, regulatory status, and human exposure. The RTECS is maintained by NIOSH and contains information on the health effects for more than 90,000 chemicals.

- Chemtrec (Chemical Transportation Emergency Center: CHEMTREC is provided by the American Chemistry Council (formerly known as the Chemical Manufacturer's Association) and provides information and assistance for emergency incidents involving chemicals and hazardous materials. CHEMTREC also supplies basic information about the production, shipping, and use of chemicals and provides information about medical treatment in response to chemical exposures.
- Material Safety Data Sheets (MSDS) are available on the Occupational Safety and Health Administration web site. MSDSs provide information such as physical and chemical properties of a substance, first aid information, emergency response, and disposal information.
- Hazardous Substances and Health Effects Database (HazDat), available on ATSDR's web site contains information on hazardous substances found at National Priorities List (NPL) and non-NPL waste sites and emergency events, and on the levels at which health effects from exposure to hazardous substances have been reported in humans and animals. HAZDAT contains environmental contamination and other data on more than 3,000 uncontrolled hazardous waste sites for which ATSDR has conducted public health assessments, prepared health consultations, or provided responses to emergencies involving releases of toxic substances into community environments. It contains toxicity information taken from the ATSDR's Toxicological Profiles for more than 200 substances most frequently found at sites.

2

Role of Enzymes in Biochemical Reactions

A living system controls its activity through enzymes. An enzyme is a protein molecule that is a biological catalyst with three characteristics. First, the basic function of an enzyme is to increase the rate of a reaction. Most cellular reactions occur about a million times faster than they would in the absence of an enzyme. Second, most enzymes act specifically with only one reactant (called a substrate) to produce products. The third and most remarkable characteristic is that enzymes are regulated from a state of low activity to high activity and vice versa. Gradually, you will appreciate that the individuality of a living cell is due in large part to the unique set of some 3,000 enzymes that it is genetically programmed to produce.

If even one enzyme is missing or defective, the results can be disastrous. Much of the information about enzymes has been made possible because they can be isolated from cells and made to work in a test tube environment.

Extensive work has also been done with X-Ray diffraction techniques to elucidate the three-dimensional structure of some enzymes. The ribbon and backbone form of carboxypeptidase is shown on the left. The substrate is shown in magenta.

Enzyme Parts List

The activity of an enzyme depends, at the minimum, on a specific protein chain. In many cases, the enzyme consists of the protein and a combination of one or more parts called cofactors. This enzyme complex is usually simply referred to simply as the enzyme.

Apoenzyme: The polypeptide or protein part of the enzyme is called the apoenzyme and may be inactive in its original synthesized structure. The inactive form of the apoenzyme is known as a proenzyme or zymogen. The proenzyme may contain several extra amino acids in the protein which are removed, and allows the final specific tertiary structure to be formed before it is activated as an apoenzyme.

Cofactors: A cofactor is a non-protein substance which may be organic, and called a coenzyme. The coenzyme is often derived from a vitamin with specific examples discussed later. Another type of cofactor is an inorganic metal ion called a metal ion activator. The inorganic metal ions may be bonded through coordinate covalent bonds. The major reason for the nutritional requirement for minerals is to supply such metal ions as Zn^{+2}, Mg^{+2}, Mn^{+2}, Fe^{+2}, Cu^{+2}, K^{+1}, and Na^{+1} for use in enzymes as cofactors.

Final Enzyme: The type of association between the cofactor and the apoenzymes varies. In some cases, the bonds are rather loose and both come together only during the course of a reaction. In other cases, they are firmly bound together by covalent bonds. The activating role of a cofactor is to either: activate the protein by changing its geometric shape, or by actually participating in the overall reaction.

The overall enzyme contains a specific geometric shape called the active site where the reaction takes place. The molecule acted upon is called the substrate.

BIOCHEMICAL FEATURES

Laccases are typically monomeric extracellular enzymes containing four copper atoms bound to 3 redox sites (T1, T2 and T3). The termed "blue copper" at the T1 site-because of its greenish-blue colour in its oxidized resting state-is responsible of the oxidation of the reducing substrate. The trinuclear cluster (containing one Cu T2 and two Cu T3) is located approx. 12 Å away from the T1 site, and it is the place where molecular oxygen is reduced to water. Laccases catalyze one-electron substrate oxidation coupled to the four-electron reduction of O_2.

It is assumed that laccases operate as a battery, storing electrons from the four individual oxidation reactions of four molecules of substrate, in order to reduce molecular oxygen to two molecules of water. Fungal laccases often occur as multiple isoenzymes expressed under different cultivation conditions (e.g. inducible or constitutive isoforms).

Most are monomeric proteins, although laccases formed by several units have been also described. They are glycoproteins with average molecular mass of 60–70 kDa, and carbohydrate contents of 10–20% which may contribute to the high stability of laccases. The covalently linked carbohydrate moiety of the enzyme is typically formed by mannose, N-acetylglucosamine and galactose. The amino acid chain contains about 520–550 amino acids including a N-terminal secretion peptide.

Biological Functions and Industrial Applications

Biological functions attributed to laccases include spore resistance and pigmentation, lignification of plant cell walls, lignin biodegradation, humus turnover and detoxification processes, virulence factors, and copper and iron homeostasis.

Laccases exhibit an extraordinary natural substrate range (phenols, polyphenols, anilines, aryl diamines, methoxy-substituted phenols, hydroxyindols, benzenethiols, inorganic/organic metal compounds and many others) which is the major reason for their attractiveness for dozens of biotechnological applications. Moreover, in the presence of small molecules, known as redox mediators, laccases enhance their substrate specificity. Indeed, laccase oxidizes the mediator and the generated radical oxidizes the substrate by mechanisms different from the enzymatic one, enabling the oxidative transformation of substrates with high redox potentials-otherwise not oxidized by the enzyme. The industrial applicability of laccase may therefore be extended by the use of a laccase-mediator system (LMS). Thus, laccase and LMS find potential application in delignification and biobleaching of pulp; treatment of wastewater from industrial plants; enzymatic modification of fibres and dye-bleaching in the textile and dye industries; enzymatic crosslinking of lignin-based materials to produce medium density fibreboard; detoxification of pollutants and bioremediation; detoxification of lignocellulose hydrolysates for ethanol production by yeast; enzymatic removal of phenolic compounds in beverages-wine and beer stabilization, fruit juice processing-; and construction of biosensors and biofuel cells.

In organic synthesis, laccases have been employed for the oxidation of functional groups, the coupling of phenols and steroids, the construction of carbon-nitrogen bonds and in the synthesis of complex natural products and more.

As mentioned above, many of these applications require the use of redox mediators opening a big window for new biotransformations of non-natural substrates towards which laccase alone hardly shows activity. On the other hand, in most of the cases large quantities of enzymes are required, which makes the efficient expression of laccase in heterologous systems an important issue.

Moreover, the protein engineering of fungal laccases with the aim of improving several enzymatic features (such as activity towards new substrates, stability under harsh operating conditions -*e.g.* presence of organic cosolvents, extreme pH values-, thermostability, and others) is a critical point in the successful application of this remarkable biocatalyst. All these issues are addressed in the following lines, paying special attention to their application in organic synthesis.

ENZYME NOMENCLATURE AND CLASSIFICATION

Enzymes are commonly named by adding a suffix "-ase" to the root name of the substrate molecule it is acting upon. For example, Lipase catalyzes the hydrolysis of a lipid triglyceride. Sucrase catalyzes the hydrolysis of sucrose into glucose and fructose.

A few enzymes discovered before this naming system was devised are known by common names. Examples are pepsin, trypsin, and chymotrypsin which catalyzes the hydrolysis of proteins. The latest systematic nomenclature system known as the International Enzyme Commission (IEC) system is based upon the type of reaction catalysed. There are six broad groups of enzymes in this system as shown in table on the left.

For example, when using this system, "urease" becomes "urea amidohydrolase." Do not be overly concerned about enzyme names, but be able to recognize a substance as an enzyme by its "-ase" ending. Some types of reactions which are being catalysed will be self evident.

Structures and Mechanisms

Enzymes are generally globular proteins and range from just 62 amino acid residues in size, for the monomer of 4-oxalocrotonate tautomerase, to over 2,500 residues in the animal fatty acid synthase. A small number of RNA-based biological catalysts exist, with the most common being the ribosome; these are referred to as either RNA-enzymes or ribozymes. The activities of enzymes are determined by their three-dimensional structure. However, although structure does determine function, predicting a novel enzyme's activity just from its structure is a very difficult problem that has not yet been solved.

Most enzymes are much larger than the substrates they act on, and only a small portion of the enzyme (around 3–4 amino acids) is directly involved in catalysis. The region that contains these catalytic residues, binds the substrate, and then carries out the reaction is known as the active site. Enzymes can also contain sites that bind cofactors, which are needed for catalysis. Some enzymes also have binding sites for small molecules, which are often direct or indirect products or substrates of the reaction catalysed. This binding can serve to increase or decrease the enzyme's activity, providing a means for feedback regulation.

Like all proteins, enzymes are long, linear chains of amino acids that fold to produce a three-dimensional product. Each unique amino acid sequence produces a specific structure, which has unique properties. Individual protein chains may sometimes group together to form a protein complex. Most enzymes can be denatured—that is, unfolded and inactivated—by heating or chemical denaturants, which disrupt the three-dimensional structure of the protein. Depending on the enzyme, denaturation may be reversible or irreversible.

Structures of enzymes in complex with substrates or substrate analogs during a reaction may be obtained using Time resolved crystallography methods.

Specificity

Enzymes are usually very specific as to which reactions they catalyse and the substrates that are involved in these reactions. Complementary shape, charge and hydrophilic/hydrophobic characteristics of enzymes and substrates are responsible for this specificity. Enzymes can also show impressive levels of stereospecificity, regioselectivity and chemoselectivity. Some of the enzymes showing the highest specificity and accuracy are involved in the copying and expression of the genome. These enzymes have "proof-reading" mechanisms. Here, an enzyme such as DNA polymerase catalyzes a reaction in a first step and then checks that the product is correct in a second step. This two-step process results in average error rates of less than 1 error in 100 million reactions in high-fidelity mammalian polymerases. Similar proofreading mechanisms are also found in RNA polymerase, aminoacyl tRNA synthetases and ribosomes.

Some enzymes that produce secondary metabolites are described as promiscuous, as they can act on a relatively broad range of different substrates. It has been suggested that this broad substrate specificity is important for the evolution of new biosynthetic pathways.

"Lock and Key" Model

Enzymes are very specific, and it was suggested by the Nobel laureate organic chemist Emil Fischer in 1894 that this was because both the enzyme and the substrate possess specific complementary geometric shapes that fit exactly into one another. This is often referred to as "the lock and key" model. However, while this model explains enzyme specificity, it fails to explain the stabilization of the transition state that enzymes achieve.

In 1958, Daniel Koshland suggested a modification to the lock and key model: since enzymes are rather flexible structures, the active site is continually reshaped by interactions with the substrate as the substrate interacts with the enzyme. As a result, the substrate does not simply bind to a rigid active site; the amino acid side chains which make up the active site are molded into the precise positions that enable the enzyme to perform its catalytic function.

In some cases, such as glycosidases, the substrate molecule also changes shape slightly as it enters the active site. The active site continues to change until the substrate is completely bound, at which point the final shape and charge is determined. Induced fit may enhance the fidelity of molecular recognition in the presence of competition and noise via the conformational proofreading mechanism.

Mechanisms

Enzymes can act in several ways, all of which lower DG‡:

- Lowering the activation energy by creating an environment in which the transition state is stabilized (e.g. straining the shape of a substrate—by binding the transition-state conformation of the substrate/product molecules, the enzyme distorts the bound substrate(s) into their transition state form, thereby reducing the amount of energy required to complete the transition).
- Lowering the energy of the transition state, but without distorting the substrate, by creating an environment with the opposite charge distribution to that of the transition state.
- Providing an alternative pathway. For example, temporarily reacting with the substrate to form an intermediate ES complex, which would be impossible in the absence of the enzyme.
- Reducing the reaction entropy change by bringing substrates together in the correct orientation to react. Considering DH‡ alone overlooks this effect.
- Increases in temperatures speed up reactions. Thus, temperature increases help the enzyme function and develop the end product even faster. However, if heated too much, the enzyme's shape deteriorates and only when the temperature comes back to normal does the enzyme regain its shape. Some enzymes like thermolabile enzymes work best at low temperatures.

Interestingly, this entropic effect involves destabilization of the ground state, and its contribution to catalysis is relatively small.

Transition State Stabilization

The understanding of the origin of the reduction of DG‡ requires one to find out how the enzymes can stabilize its transition state more than the transition state of the uncatalyzed reaction.

Apparently, the most effective way for reaching large stabilization is the use of electrostatic effects, in particular, by having a relatively fixed polar environment that is oriented toward the charge distribution of the transition state. Such an environment does not exist in the uncatalyzed reaction in water.

Dynamics and Function

The internal dynamics of enzymes is linked to their mechanism of catalysis. Internal dynamics are the movement of parts of the enzyme's structure, such as individual amino acid residues, a group of amino acids, or even an entire protein domain. These movements occur at various time-scales ranging from femtoseconds to seconds. Networks of protein residues throughout an enzyme's structure can contribute to catalysis through dynamic motions. Protein motions are vital to many enzymes, but whether small and fast vibrations, or larger and slower conformational movements are more important depends on the type of reaction involved. However, although these movements are important in binding and releasing substrates and products, it is not clear if protein movements help to accelerate the chemical steps in enzymatic reactions. These new insights also have implications in understanding allosteric effects and developing new drugs.

Allosteric Modulation

Allosteric sites are sites on the enzyme that bind to molecules in the cellular environment. The sites form weak, noncovalent bonds with these molecules, causing a change in the conformation of the enzyme. This change in conformation translates to the active site, which then affects the reaction rate of the enzyme. Allosteric interactions can both inhibit and activate enzymes and are a common way that enzymes are controlled in the body.

BIOCHEMICAL OR ENZYMATIC PATHWAYS

Sometimes a cell needs a product from a reactant that requires only a single chemical reaction. For example, reactant A needs to be changed to product B. This reaction would only require the enzyme that fits with its substrate, reactant A. However, this is not usually the case. Most products inside cells that need to be broken down or built up require a multi-step process. This process is called a biochemical pathway. It goes something like this:

- Enzyme A changes substrate A to product B.
- Product or substrate B now can be bound by enzyme B and be changed to product or substrate C
- Substrate C now fits with enzyme C which changes it to product D.
- If product D is what the cell needs, the pathway is finished. If not, the pathway can continue in the same way until the ultimate product is obtained.

Finally a quick lesson in naming enzymes. Generally to name an enzyme we take the substrate it breaks down or product that it manufactures and add an -ase on the end.

Examples: enzymes that degrade the following substances DNAse, ATPase or enzymes that build up the following substances: DNA polymerase and ATP synthetase.

BIOCHEMICAL PROPERTIES

Denaturation

The strands of the DNA double helix are held together by hydrogen bonding interactions between the complementary base pairs. Heating DNA in solution easily breaks these hydrogen bonds, allowing the two strands to separate—a process called denaturation or melting. The two strands may reassociate when the solution cools, reforming the starting DNA duplex—a process called renaturation or hybridization. These processes form the basis of many important techniques for manipulating DNA. For example, a short piece of DNA called an oligonucleotide can be used to test whether a very long DNA sequence has the complementary sequence of the oligonucleotide embedded within it. Using hybridization, a single-stranded DNA molecule can capture complementary sequences from any source. Single strands from RNA can also reassociate. DNA and RNA single strands can form hybrid molecules that are even more stable than double-stranded DNA. These molecules form the basis of a technique that is used to purify and characterize messenger RNA (mRNA) molecules corresponding to single genes.

Ultraviolet Absorption

DNA melting and reassociation can be monitored by measuring the absorption of ultraviolet (UV) light at a wavelength of 260 nanometres (billionths of a metre). When DNA is in a double-stranded conformation, absorption is fairly weak, but when DNA is single-stranded, the unstacking of the bases leads to an enhancement of absorption called hyperchromicity. Therefore, the extent to which DNA is single-stranded or double-stranded can be determined by monitoring UV absorption.

Chemical Modification

After a DNA molecule has been assembled, it may be chemically modified—sometimes deliberately by special enzymes called DNA methyltransferases and sometimes accidentally by oxidation, ionizing radiation, or the action of chemical carcinogens. DNA can also be cleaved and degraded by enzymes called nucleases.

Mutation

Chemical modification of DNA can lead to mutations in the genetic material. Anions such as bisulfite can deaminate cytosine to form uracil, changing the genetic message by causing C-to-T transitions. Exposure to acid causes the loss of purine residues, though specific enzymes exist in cells to repair these lesions. Exposure to UV light can cause adjacent pyrimidines to dimerize, while oxidative damage from free radicals or strong oxidizing agents can cause a variety of lesions that are mutagenic if not repaired. Halogens such as chlorine and bromine react directly with uracil, adenine, and guanine, giving substituted bases that are often mutagenic. Similarly, nitrous acid reacts with primary amine groups—for example, converting adenosine into inosine—which then leads to changes in base pairing and mutation. Many chemical mutagens, such as chlorinated hydrocarbons and nitrites, owe their toxicity to the production of halides and nitrous acid during their metabolism in the body.

Supercoiling

Circular DNA molecules such as those found in plasmids or bacterial chromosomes can adopt many different topologies. One is active supercoiling, which involves the cleavage of one DNA strand, its winding one or more turns around the complementary strand, and then the resealing of the molecule. Each complete rotation leads to the introduction of one supercoiled turn in the DNA, a process that can continue until the DNA is fully wound and collapses on itself in a tight ball. Reversal is also possible. Special enzymes called gyrases and topoisomerases catalyse the winding and relaxation of supercoiled DNA. In the linear chromosomes of eukaryotes, the DNA is usually tightly constrained at various points by proteins, allowing the intervening stretches to be supercoiled. This property is partially responsible for the great compaction of DNA that is necessary to fit it within the confines of the cell. The DNA in one human cell would have an extended length of between two and three metres, but it is packed very tightly so that it can fit within a human cell nucleus that is 10 micrometres in diameter.

CARBON-BASED ALTERNATIVES TO HYDROCARBONS

In hydrogen-depleted or strongly oxidizing environments, a carbon-based biology in which little to no hydrogen is used might be a possibility. Such "hydrogenless life" would use oxocarbons (such as mellitic anhydride or similar) as its building blocks, instead of hydrocarbons.

Chlorine as an Alternative to Oxygen

A number of alternatives to molecular oxygen as a terminal electron acceptor are known from anaerobic life forms on Earth. However, it has been proposed that chlorine might serve as a more general biological alternative to oxygen, either in carbon-based biologies or hypothetical non-carbon-based ones. But chlorine is much less abundant than oxygen in the universe, and so planets with a sufficiently chlorine-rich atmosphere are likely to be rare, if they exist at all. Chlorine will instead likely be bound up as salts and other inert compounds.

Arsenic as an Alternative to Phosphorus

Arsenic, which is chemically similar to phosphorus, while poisonous for most life forms on Earth, is incorporated into the biochemistry of some organisms. Some marine algae incorporate arsenic into complex organic molecules such as arsenosugars and arsenobetaines. Fungi and bacteria can produce volatile methylated arsenic compounds. Arsenate reduction and arsenite oxidation have been observed in microbes (*Chrysiogenes arsenatis*). Additionally, some prokaryotes can use arsenate as a terminal electron acceptor during anaerobic growth and some can utilize arsenite as an electron donor to generate energy.

It has been speculated that the earliest life forms on Earth may have used arsenic in place of phosphorus in the backbone of their DNA. A common objection to this scenario is that arsenate esters are so much less stable to hydrolysis than corresponding phosphate esters that arsenic would not be suitable for this function. The authors of a 2010 geomicrobiology study supported in part by NASA have postulated that a bacterium, named GFAJ-1, collected in the sediments of Mono Lake in eastern California, can employ such 'arsenic DNA' when cultured without phosphorus. They proposed that the bacterium may employ high levels of poly-â-hydroxybutyrate or other means to reduce the effective concentration of water and stabilize its arsenate esters.

This claim was heavily criticized almost immediately after publication for the perceived lack of appropriate controls Science writer Carl Zimmer contacted several scientists for an assessment: "I reached out to a dozen experts... Almost unanimously, they think the NASA scientists have failed to make their case".

Selenium or Tellurium as an Alternative to Sulfur

Some organisms are already known to feature selenoproteins, in which sulfur is replaced by selenium. Some fungi also can produce telluro-methionine and telluro-cysteine. No organism has yet been shown to *completely* replace sulfur with the heavier chalcogens, however.

Non-water Solvents

In addition to carbon compounds, all currently known terrestrial life also requires water as a solvent. It is sometimes assumed that water is the only suitable chemical to fill this role. Some of the properties of water that are important for life processes include a large temperature range over which it is liquid, a high heat capacity useful for temperature regulation, a large heat of vaporization, and the ability to dissolve a wide variety of compounds. Water is also amphoteric, meaning it can donate and accept a proton allowing it to act as an acid or a base. This property is crucial in many organic and biochemical reactions, where water serves as a solvent, a reactant, or a product. There are other chemicals with similar properties that have sometimes been proposed as alternatives. Additionally, water is the only compound listed here that is less dense as a solid (ice) than as a liquid. This is why bodies of water freeze over but do not freeze solid (from the bottom up). If ice were denser than liquid water (as is true for nearly all other compounds) then large bodies of liquid would slowly freeze solid, which would not be conducive to the formation of life.

There are some properties that make certain compounds and elements much more favourable than others as solvents in a successful biosphere. The first of these is a reasonably wide temperature range in which it is liquid. Using ammonia as a baseline, the boiling point must be at least 20% greater than the melting point measured in Kelvin. For example, the boiling point of water is 373.15 K (at 1 atmosphere), while the melting point is 273.15 K. 373.15/273.15=1.37, meaning that the boiling point is about 37% higher than the freezing point, and thus water has an acceptable temperature range. This means that polar solvents, like water (37%), tend to do better than nonpolar solvents, such as methane (23%). However, as shown in the example given, methane does pass the test, albeit by a smaller margin than water. Since boiling points can vary based on the pressure, the question tends not to be *does* the prospective solvent pass this test, but *at what pressure* does the prospective solvent pass this test. For example, hydrogen cyanide does not pass this test at 1 atmosphere, with 15%, but in a Venus pressure atmosphere, with 92 bars (9.2 MPa) of pressure, it is very likely that it would.

Another important characteristic of a prospective solvent is that one can reasonably expect to find it in the location that the theoretical life will live. Usually, this is a planet or moon, but other locations such as a planetary nebula or a gas giant are also possible. This requirement has two main consequences. The first of these is that it needs to be composed of common elements. On a planet or moon, the most common elements tend to be carbon, hydrogen, oxygen, nitrogen, silicon, sulfur, and various metals. Solvents made of a combination of these elements tend to have an advantage. This is why hydrogen fluoride, shown below, is not considered a likely solvent. The second offshoot of that is that the prospective solvent needs to be something that can reasonably be expected to be synthesized from these elements in large quantities, using solely natural phenomena. For example, the compounds methane, ammonia, and water, while not only being on the list of common elements, are also widespread, and thus one can probably count on their presence in the relevant temperature ranges. However, while aluminium fits the liquid range requirements (its number is about 200% at 1 atmosphere), and is common (it is about 7% of the composition of rocky bodies), it is almost universally found in the form of its oxide, aluminium oxide, because not even carbon can reduce it chemically. This also means that a prospective solvent cannot be overly reactive, in its given atmosphere, for example sodium because then it would probably not remain in existence.

Ammonia

Ammonia is perhaps the most commonly proposed alternative. Numerous chemical reactions are possible in an ammonia solution, and liquid ammonia has some chemical similarities with water. Ammonia can dissolve most organic molecules at least as well as water does, and in addition it is capable of dissolving many elemental metals. Given this set of chemical properties it has been theorized that ammonia-based life forms might be possible.

However, ammonia has some problems as a basis for life. The hydrogen bonds between ammonia molecules are weaker than those in water, causing ammonia's heat of vaporization to be half that of water, its surface tension to be three times smaller, and reducing its ability to concentrate non-polar molecules through a hydrophobic effect. For these reasons, some members of the scientific community question how well ammonia could hold prebiotic molecules together to allow the emergence of a self-reproducing system. Ammonia is also flammable and can be oxidized and could not exist sustainably in a biosphere that oxidizes it. It would, however, be stable in a reducing environment.

A biosphere based on ammonia would likely exist at temperatures or air pressures that are extremely unusual for terrestrial life. Terrestrial life usually exists within the melting point and boiling point of water at normal pressure, between 0 °C (273 K) and 100 °C (373 K); at normal pressure ammonia's melting and boiling points are between H78 °C (195 K) and H33 °C (240 K). Such extremely cold temperatures create problems, as they slow biochemical reactions tremendously and may cause biochemical precipitation out of solution due to high melting points. Ammonia could be a liquid at normal temperatures, but at much higher pressures; for example, at 60 atm, ammonia melts at H77 °C (196 K) and boils at 98 °C (371 K).

Ammonia and ammonia-water mixtures remain liquid at temperatures far below the freezing point of pure water, so such biochemistries might be well suited to planets and moons orbiting outside the water-based habitability zone. Such conditions could exist, for example, under the surface of Saturn's largest moon Titan.

Methane

A mixture of hydrocarbons, such as the methane/ethane lakes detected on Titan by the Cassini spacecraft, could act as a solvent over a wide range of temperatures but would lack polarity. There is debate about the effectiveness of methane as a medium for life compared to water or ammonia. While water is a far better solvent than methane, enabling easier transport of substances in a cell, methane's lesser chemical reactivity allows for the easier formation of large structures akin to proteins. Isaac Asimov, the biochemist and science fiction writer, suggested that poly-lipids could form a substitute for proteins in a non-polar solvent such as methane or liquid hydrogen.

Possible evidence for this form of life on Titan was identified in 2010 by Darrell Strobel of Johns Hopkins University; an over-abundance of molecular hydrogen in Titan's upper atmospheric layers, which leads to a downward flow at a rate of roughly 10^{25} molecules per second. Near the surface the hydrogen apparently disappears, which may imply its consumption by methanogenic life forms. Another paper released the same month showed little evidence of acetylene on Titan's surface, where scientists had expected the compound to accumulate; according Strobel, this is consistent with the hypothesis that acetylene is being consumed by methanogens. Chris McKay, while agreeing that presence of life is a possible explanation for the findings about hydrogen and acetylene, has cautioned that other explanations are currently more likely: namely the possibility that the results are due to human error, or to the presence of some as-yet unknown catalyst in the soil. He noted that such a catalyst, effective at 95 Kelvin, would in itself be a startling discovery.

Hydrogen Fluoride

Hydrogen fluoride, like water, is a polar molecule, and due to its polarity it can dissolve many ionic compounds. Its melting point is H84 °C and its boiling point is 19.54 °C (at atmospheric pressure); the difference between the two is little more than 100 °C. HF also makes hydrogen bonds with its neighbor molecules as do water and ammonia. These would make HF a candidate to host life on other planets.

Not much research has been done on liquid HF in regards to its ability to dissolve and react with non-polar molecules. The biota in an HF ocean could use the fluorine as an electron acceptor to photosynthesize energy. HF is dangerous to the systems of molecules that earth-life is made of, but the paraffins are stable with it. But presence of great amounts of free HF in a planetary scale, like water on Earth, is doubtful, even next to impossible: water would be present in it, forming a solution of hydrofluoric acid that would change all its properties. The hypothetical planet would also surely contain silicates that would react with HF to form inert compounds as silicon fluorides, as soon as HF would be present; thus preventing its concentration in great quantities in a hypothetical planetary environment. The cosmic abundance of fluorine is low, and it forms chemically inert compounds rapidly in interstellar nebulae, because it is the most reactive element.

Other Solvents or Cosolvents

Other solvents sometimes proposed include formamide, methanol, hydrogen sulfide and hydrogen chloride. Hydrogen chloride suffers from the low cosmic abundance of chlorine, while hydrogen sulfide suffers from its high reactivity. Moreover, the first two could not be expected to be found in vast quantities on a planetary scale, and would only be part of the internal physiology of organisms.

A proposal has been made that life on Mars may exist and be using a mixture of water and hydrogen peroxide as its solvent. A 61.2% (by weight) mix of water and hydrogen peroxide has a freezing point of H56.5 °C, and also tends to super-cool rather than crystallize. It is also hygroscopic, an advantage in a water-scarce environment.

Molten Salts

At extremely high temperatures, water and most other molecular compounds will have already evaporated. A molten salt, or a eutectic mixture of such salts, would remain liquid and could work as a solvent. Life in such circumstances would have to be based on refractory materials like metal oxides. But it's very doubful that metal oxide compounds could take part in a biochemistry, being quite simple and inert compounds.

Other Types of Speculations

Non-green Photosynthesizers

Physicists have noted that, while photosynthesis on Earth generally involves green plants, a variety of other colored plants could also support photosynthesis, essential for most life on Earth, and that other colors might be preferred in places that receive a different mix of stellar radiation than that received on Earth. These studies indicate that while blue photosynthetic plants would be unlikely (because absorbed blue light provides some of the highest photosynthetic yields in the light spectrum), yellow or red plants are plausible. These conclusions are based in part on the luminosity spectra of different types of stars, the transmission characteristics of hypothetical planetary atmospheres, and the absorption spectra of various photosynthetic pigments from organisms on Earth. The first plants on Earth may have been a slightly different colour, because the Sun was, in the first geological eons, a little less bright, and its light was filtered by passing through an atmosphere with a different composition.

Alternative Atmospheres

The gases present in the atmosphere on Earth have varied greatly over its history. Traditional plant photosynthesis has terraformed the atmosphere by sequestering carbon from carbon dioxide, increasing the proportion of molecular oxygen, and by participating in the nitrogen cycle. Modern oxygen breathing animals would have been biochemically impossible until earlier photosynthetic life transformed Earth's atmosphere. The first dramatic rise in atmospheric oxygen on Earth, to about a tenth of its present-day value, occurred approximately 2.5 billion years ago, and that level did not change significantly until the Cambrian era approximately 600 million years ago.

Changes in the gas mixture in the atmosphere, even in an atmosphere made up predominantly of the same molecules of Earth's atmosphere, impacts the biochemistry and morphology of life. For example, periods of high oxygen concentrations determined from ice core samples have been associated with fauna of a larger scale in the fossil record, while periods associated with of low oxygen concentrations have been associated with fauna of a smaller scale in the fossil record.

Also, while it is customary to think of plants on one side of the oxygen and nitrogen cycles as being sessile, and of animals on the other side as being motile, this is not a biological imperative. There are animals which are sessile for all or most of their lives (such as corals), and there are plants (such as tumbleweeds, and venus fly traps) that exhibit more mobility than is customarily associated with plants. On a slowly rotating planet, for example, it might be adaptive for photosynthesis to be performed by "plants" that can move to remain in the light, like Earth's sunflowers; while non-photosynthetic "animals", much like Earth's fungi, might have a lesser need to move from place to place on their own. This would be a mirror image of Earth's ecology.

Variable Environments

Many Earth plants and animals undergo major biochemical changes during their life cycles as a response to changing environmental conditions, for example, by having a spore or hibernation state that can be sustained for years or even millennia between more active life stages. Thus, it would be biochemically possible to sustain life in environments that are only periodically consistent with life as we know it.

For example, frogs in cold climates can survive for extended periods of time with most of their body water in a frozen state, while desert frogs in Australia can become inactive and dehydrate in dry periods, losing up to 75% of their fluids, yet return to life by rapidly rehydrating in wet periods. Either type of frog would appear biochemically inactive (i.e. not living) during dormant periods to anyone lacking a sensitive means of detecting low levels of metabolism.

Nonplanetary life

Dust and Plasma-based

In 2007 V. N. Tsytovich and colleagues proposed that life-like behaviors could be exhibited by dust particles suspended in a plasma, under conditions that might exist in space. Computer models showed that when the dust became charged the particles could self-organize into microscopic helical structures capable of replicating themselves, interacting with other neighboring structures, and evolving into more stable forms. Similar forms of life were described in Fred Hoyle's classic novel *The Black Cloud*.

Usage in Fiction

In the realm of science fiction, there have occasionally been forms of life proposed that, while often highly speculative and unsupported by rigorous theoretical examination, are nevertheless interesting and in some cases even plausible.

Novels, Short Stories and Comics

In Arthur C. Clarke's short story "Technical Error" there is an example of differing chirality. This is not a case of alien life, rather it is an accident. A technical error in a new power plant changes the chirality of a worker. He cannot read because he sees all type as reversed, and he believes his left side is his right. The amusement ends when it is discovered that he is dying from malnutrition because his new biology cannot utilize normal food.

The concept of reversed chirality also figured prominently in the plot of James Blish's *Star Trek* novel *Spock Must Die!*, where a transporter experiment gone awry ends up creating a duplicate Spock who turns out to be a perfect mirror-image of the original all the way down to the atomic level. An example of silicon based life forms takes place in the Alan Dean Foster novel *Sentenced to Prism* in which the protagonist Evan Orgell is trapped on a planet whose entire ecosystem is mostly silicon-based. Perhaps the most extreme example in science fiction is James White's *Sector General*: a series of novels and short stories about multienvironment hospital for the strangest life-forms imaginable, some of them breathing methane, chlorine, water and sometimes also oxygen. Some of the species *metabolise* directly hard radiation and their environment doesn't differ much from the atmosphere of a star, while others live in near absolute zero temperatures. All life forms are classified according to their metabolism, internal and external features, and more extreme abilities (telepathy, empathy, hive mind, etc.) with four letter codes. Humans from Earth share the *DBDG* specification with small furry beings called Nidians.

One of the major sentient species in Terry Pratchett's *Discworld* universe are the "earth"-based (ranging from Detritus to Diamond) Trolls. Pratchett has also written the science fiction novel *The Dark Side of the Sun* which features a range of extraordinary life-forms, including a telepathic body of water, creatures called "Sundogs", which are capable of interstellar travel from birth, and a sentient planet: effectively a giant silicon-based computer. Fred Hoyle's classic novel *The Black Cloud* features a life form consisting of a vast cloud of interstellar dust, the individual particles of which interact via electromagnetic signalling analogous to how the individual cells of multicellular terrestrial life interact. Outside of science-fiction, life in interstellar dust has been proposed as part of the panspermia hypothesis. The low temperatures and densities of interstellar clouds would seem to imply that life processes would operate much more slowly there than on Earth. Inorganic dust-based life has been speculated upon based on recent computer simulations.

Similarly, Arthur C. Clarke's "Crusade" revolves around a planetwide life-form based on silicon and superfluid helium located in deep intergalactic space, processing its thoughts slowly by human standards, that sends probes to look for similar life in nearby galaxies. It concludes that it needs to make planets more habitable for similar life-forms, and sends out other probes to foment supernovae to do so. Clarke implies that this is what accounts for most supernovae having occurred in the same region of space and warns that the effort will eventually reach Earth.

Robert L. Forward's *Camelot 30K* describes an ecosystem on the surface of Kuiper belt objects that is based on a fluorocarbon chemistry with OF_2 as the principal solvent instead of H_2O. The organisms in this ecology keep warm by secreting a pellet of uranium-235 inside themselves and then moderating its nuclear fission using a boron-rich carapace around it. Kuiper belt objects are known to be rich in organic compounds such as tholins, so some form of life existing on their surfaces is not entirely implausible–though perhaps not going so far as to develop natural internal nuclear reactors, as have Forward's. Fluorine is also of low cosmic abundance, so its use in this manner is unlikely.

In Forward's *Rocheworld* series, an Earth-like biochemistry is proposed that uses a mixture of water and ammonia as its solvent. In *Dragon's Egg* and *Starquake,* Forward proposes life on the surface of a neutron star utilizing "nuclear chemistry" in the degenerate matter crust. Since such life utilised strong nuclear forces instead of electromagnetic interactions, it was posited that life might function millions of times faster than typical on Earth.

Gregory Benford and David Brin's *Heart of the Comet* features a comet with a conventional carbon-and-water-based ecosystem that becomes active near the perihelion when the Sun warms it. Brin's own novel *Sundiver* is an example of science fiction proposing a form of life existing within the plasma atmosphere of a star using complex self-sustaining magnetic fields. Similar sorts of plasmoid life have sometimes been proposed to exist in other places, such as planetary ionospheres or the interstellar medium, but usually only by fringe theorists. Gregory Benford had a form of plasma-based life exist in the accretion disk of a primordial black hole in his novel *Eater*.

The suggestion that life could even occur within the plasma of a star has been picked up by other science fiction writers, as in David Brin's Uplift Saga. The idea is that places where reactions occur–even an incredible environment as a star–presents a possible medium for some chain of events that could produce a system able to replicate.

The Outsiders in Larry Niven's Known Space universe are cryogenic creatures based on liquid helium. They derive thermoelectric energy from a temperature gradient by basking half their body in sunlight, keeping the other half in shadow and exposed to interstellar vacuum.

Stephen Baxter has imagined perhaps some of the most unusual exotic life-forms in his Xeelee series of novels and stories, including supersymmetric photino-based life that congregate in the gravity wells of stars, entities composed of quantum wave functions, and the Qax, who thrive in any form of convection cells, from swamp gas to the atmospheres of gas giants. In his book *Manifold: Space,* he also proposes natural robots, life forms made of iron, called the Gaijin, evolving from creatures in oceans of iron carbonyl.

In his novel *Diaspora,* Greg Egan posits entire virtual universes implemented on Turing Machines encoded by Wang Tiles in gargantuan polysaccharide 'carpets.' The sentient ocean that covers much of the surface of Solaris in Stanislaw Lem's eponymous novel also seems, from much of the fictional research quoted and discussed in the book, to be based on some element other than carbon. In the same novel Egan describes lifeforms in the 6-D 'macrosphere' which use a collapsed atom chemistry with energetic processes of the same order as nuclear reactions, due to the peculiarities of higher dimensional physics.

In her novel *Brain Plague*, Joan Slonczewski describes a species of intelligent microrganisms with arsenic based chemistries that live symbiotically with human hosts. Sergeant Schlock is one of the lead characters in the webcomic *Schlock Mercenary*. His species, Carbosilicate Amorphs, evolved from self-repairing distributed data storage devices, and as such, redundantly distribute their 'brain' throughout their body. They are highly resistant to hard vacuum, explosive decompression, projectile weapons, chemical-based explosives, and dismemberment. Their only specialty organ is their eyes, which they harvest as fruit from the Ghanj-Rho eye-tree on their home planet. While the Amorphs have the ability to move fast, quietly, and sprout appendages at will, they excel at 'closer-than-melee-range combat, primarily "meme-toxins" against other Amorphs.

A more farcical example comes from *The Hitchhiker's Guide to the Galaxy*, where the Hooloovoo are a hyperintelligent shade of the colour blue. Alien warriors recruited by the god Klael in David Eddings' "Tamuli" trilogy are noted by their human opponents to breathe marsh-gas (methane). Within Eddings' universe, this limits their capacity for exertion in an oxygen atmosphere, and also determines the tactics used to fight them and eventually to destroy them in their encampments. The eponymous organism in Michael Crichton's *The Andromeda Strain* is described as reproducing via the direct conversion of energy into matter.

Star Trek

A well-known example of a non–carbon-based life-form in science fiction is the Horta in the original *Star Trek* episode "Devil in the Dark". A highly intelligent silicon-based creature made almost entirely of pure rock, it tunnels through rock as easily as humans move through air. The entire species dies out every 50,000 years except for one who tends for all the eggs, which take the form of silicon nodules scattered throughout the caverns and tunnels of its home planet, Janus VI. The inadvertent destruction of many of these eggs by a human mining colony led the mother Horta to respond by killing the colonists and sabotaging their equipment; it was only through a Vulcan mind meld that the race's benevolence and intelligence were discovered and peaceful relations established.

Star Trek would later offer other corporeal life-forms with an alternative biochemistry. The Tholians of "The Tholian Web" are depicted and described, in that episode and later in the *Star Trek: Enterprise* episode "In a Mirror, Darkly" as being primarily of mineral-based composition and thriving only in superheated conditions. Another episode from *TOS*'s third season, "The Savage Curtain", depicted another rock creature called an Excalbian, which is believed in fanon to also have been silicon-based.' In *Star Trek: The Next*

Generation, the Crystalline Entity appeared in two episodes, "Datalore" and "Silicon Avatar". This was an enormous spacefaring crystal lattice that had taken thousands of lives in its quest for energy.

It may have been unaware of this, however, but it was destroyed before communications could be established at a level sufficient to ascertain it. In another episode, "Home Soil", intelligent crystals that formed a "microbrain" were discovered during a terraforming mission, and they described the humans they encountered as "ugly bags of mostly water."

"The Disease", an episode of *Star Trek: Voyager* featured some artificially-engineered silicon-based parasites, and an *Enterprise* episode, "Observer Effect", also presented a lethal silicon-based virus. In another Voyager episode, "Hope and Fear", a xenon-based life-form was mentioned. In the *Enterprise* episode "The Communicator", an alien species is encountered whose blood chemistry, while not explicitly stated, is sufficiently different from terrestrial organisms that it is not red and iron is toxic to it.

Star Wars

In the *Star Wars* movie *The Empire Strikes Back,* two life-forms were encountered by the characters that were non-carbon based entities. Although details of their physiology were not mentioned on screen, the space slug, (a giant worm-like creature that lived on asteroids in the freezing vacuum of space), and the Mynock, (pesky bat-like vermin that would attach to spaceship hulls and chew through power conduits to feed off the raw energy), are said to be silicon-based organisms in Star Wars Expanded Universe sources. Also from *The Empire Strikes Back,* the bounty hunter Zuckuss is a member of the Gand race, an ammonia-based life-form. However, it is worth noting that the Gand are divided into two subspecies, only one of which breathes at all, the other drawing all their required sustenance from food intake and producing speech by essentially modulated flatulence.

Appearing only in the Star Wars Expanded Universe is the spice spider of Kessel, a creature made of glitterstim spice and silicon that spun crystalline webs harvested by miners as glitterstim spice, an illegal psychoactive narcotic. The spider used the webs to catch bogeys, tiny energy creatures that it consumed for energy.

Other Film and Television

- In the movie *Titan A.E.,* an alien race called the Drej were composed of a coherent plasma.
- In the movie *The Monolith Monsters,* a silicon meteor reproduces, draining silicates from everything it touches. It needs water to start its cycle and contains molecular structures typical of many kinds of

rocks, mixed together. A geologist says that its structure is nearly impossible. The meteor is killed by salt water, which can stop the cycle.

- In "Firewalker", a second-season episode of *The X-Files,* a silicon-based plant that infects humans parasitically through its spores is discovered living deep in a volcano.
- Also from *The X-Files,* the first-season episode "Ice" deals with an ammonia-based vermiform parasite.
- A key plot point in the comedy *Evolution* involves nitrogen-based life forms, and using selenium-based shampoo to poison them (with the bonus of a product placement for Head & Shoulders).
- In the *Stargate SG-1* fourth season episode "Scorched Earth", a Human society known as the Enkarans are threatened on their new homeworld by an alien ship that is terraforming the planet to be suitable for the sulfur-based Gadmeer species.
- In the *Stargate Atlantis* fifth season episode "Remnants", a device is found whose purpose was to seed a planet with silicon-based life.
- In *Ben 10,* both the Omnitrix alien Diamondhead and the alien bounty hunter Tetrax Shard are members of the Petrosapien species, which are a form of silicon-based life. Other silicon-based lifeforms on the show include the Omnitrix aliens Chromostone (who is crystaline), and Echo Echo and Upgrade (who are both biomechanical). Other member's of Upgrade's species have appeared, including the shape-shifting "Ship," a pet of Ben's girlfriend, Julie.
- In *Dragon Ball Z,* shortly before the destruction of Planet Namek, Frieza tells Goku that he does not need to breathe. This would allow him to survive the destruction of the planet and suggest that he is not a carbon-based life-form.
- *Indiana Jones and the Kingdom of the Crystal Skull* (2008) introduces thirteen "extra dimensional beings" with crystal skeletons, who founded a city that became the basis of the El Dorado myth. Though their flesh has died and rotted away, their minds still live on within their skeletons, which communicate telepathically.
- An episode ("The Chrysalids") of the series *Space: 1999* presents humanoids on a planet protected by a electric waves system, which are living in a chlorine atmosphere.
- The episode "The Stones of Blood", of the 16th season of *Doctor Who,* the Fourth Doctor encounters the Ogri, a silicon-based life form, and in the same sub-plot, the Megara, who are made entirely out of an unknown substance, possibly energy, and they uphold the word of the law, and execute all who break the law with a beam of energy.

Computer and Video Games

In the *Command & Conquer* real-time strategy games, both the gameplay and storyline revolve heavily around the introduction to Earth of an extraterrestrial mutagen called Tiberium via meteor, which displays strikingly lifelike behaviours such as self-replication, evolution, and homeostasis, without undergoing anything like common carbon-based metabolic cycles, and which appears to be colonising the Earth, converting it into an environment unsuited to carbon-based biology. Earth creatures (such as animals, plants and even humans) exposed to Tiberium can either be killed because of the radiation or be transformed into Tiberium-based life-forms, to whom Tiberium radiation is curative rather than toxic. It is later revealed that Tiberium was introduced to earth by the Scrin, an extremely advanced race of Tiberium-based aliens bent on mining the planet after the Tiberium deposits have reached maturity.

In the *Halo* video game series, a race of Covenant aliens named "Grunts" by humans require a breathing apparatus while fighting the humans in an Earth-like atmosphere. According to the novelizations of the video game, the Grunts' apparatus allows them to breathe the methane they need to survive.

In the *Master of Orion* series of space strategy games, there exists an extraterrestrial race called Silicoids, whose appearance (and presumably composition) is similar to crystalline mineral structures. The game posits that this grants them immunity to the effects of hostile environments and pollution and they require no sustenance, at the expense of impeding their reproductive rate and their ability to interact with other intelligent species.

In the *Metroid Prime* series, Phazon is a highly radioactive, self regenerating mineral with organic properties that is generated by the sentient planet Phaaze.

In *Metroid Prime Hunters,* Spire is a rock-like, silicon based alien. He is the last Diamont (presumably a play on the word diamond, which is composed of carbon).

In the *Submarine TITANS* strategy game, the alien race in the game are called "the Silicons" because they are silicon based life forms.

In the *Star Control* series, the Chenjesu, are hyperintelligent, peaceful silicon-based life-forms that were the backbone of the Alliance of Free Stars. Their crystalline biology apparently gives them the ability to send and receive hyperwave transmissions. Also, there are the Slylandro who are gas beings residing in the upper atmosphere of a gas giant. As well, there are evidences of another silicon-based race, the Taalo who are described by the xenophobic Ur-Quan as the only race to have not awakened their territorial instincts. The Taalo were also immune to mind control

In the game of *Xenosaga,* artificial life forms known as Realians have been created using silicon-based chemistry. They resemble humans in every aspect, except they are considered to be lower than humans on the social ladder.

In *Mass Effect* the alien Turians and Quarians, are both based on dextro-amino acids, as opposed to the all other sentient species of the galaxy based on levo-amino acids. There are also the Volus, an ammonia based species that must wear pressure suits to survive in environments suited to the other races.

In *Spore*, the Grox refer to the player and to other alien empires as "slow thinking carbon-based lifeforms" and "carbon wads", implying that the Grox (which are at least partly machine life) are not carbon-based. Also, the Grox can only exist on barren planets which cannot support other life, and when a planet is terraformed the Grox inhabiting it die immediately. The Grox seem to gather sustenance from the radiation from the galactic core, as the Grox colonies are larger the closer they are to the galactic core.

3

Traditional Toxicology

The traditional definition of toxicology is "the science of poisons." As our understanding of how various agents can cause harm to humans and other organisms, a more descriptive definition of toxicology is "the study of the adverse effects of chemicals or physical agents on living organisms".

These adverse effects may occur in many forms, ranging from immediate death to subtle changes not realized until months or years later. They may occur at various levels within the body, such as an organ, a type of cell, or a specific biochemical. Knowledge of how toxic agents damage the body has progressed along with medical knowledge. It is now known that various observable changes in anatomy or body functions actually result from previously unrecognized changes in specific biochemicals in the body.

The historical development of toxicology began with early cave dwellers who recognized poisonous plants and animals and used their extracts for hunting or in warfare. By 1500 BC, written recordings indicated that hemlock, opium, arrow poisons, and certain metals were used to poison enemies or for state executions.

With time, poisons became widely used and with great sophistication. Notable poisoning victims include Socrates, Cleopatra, and Claudius. By the time of the Renaissance and Age of Enlightenment, certain concepts fundamental to toxicology began to take shape.

Paracelsus determined that specific chemicals were actually responsible for the toxicity of a plant or animal poison. He also documented that the body's response to those chemicals depended on the dose received. His studies revealed that small doses of a substance might be harmless or beneficial whereas larger doses could be toxic. This is now known as the dose-response relationship, a major concept of toxicology. Paracelsus is often quoted for his statement: "All substances are poisons; there is none which is not a poison. The right dose differentiates a poison and a remedy." Orfila, a Spanish physician, is often referred to as the founder of toxicology. It was Orfila who first prepared a systematic correlation between the chemical and biological properties of poisons of the time. He demonstrated effects of poisons on specific organs by analyzing autopsy materials for poisons and their associated tissue damage.

The 20th century is marked by an advanced level of understanding of toxicology. DNA (the molecule of life) and various biochemicals that maintain body functions were discovered. Our level of knowledge of toxic effects on organs and cells is now being revealed at the molecular level. It is recognized that virtually all toxic effects are caused by changes in specific cellular molecules and biochemicals.

Xenobiotic is the general term that is used for a foreign substance taken into the body. It is derived from the Greek term xeno which means "foreigner." Xenobiotics may produce beneficial effects (such as a pharmaceuticals) or they may be toxic (such as lead). As Paracelsus proposed centuries ago, dose differentiates whether a substance will be a remedy or a poison. A xenobiotic in small amounts may be non-toxic and even beneficial but when the dose is increased, toxic and lethal effects may result.

BASIC TOXICOLOGY TERMINOLOGY

Toxicology is an evolving medical science and so is toxicology terminology. Most terms are quite specific and will be defined as they appear in the tutorial. Those most commonly used are discussed in the following frames.

Toxicology is the study of the adverse effects of chemicals or physical agents on living organisms. A toxicologist is a scientist that determines the harmful effects of agents and the cellular, biochemical, and molecular mechanisms responsible for the effects.

Terminology and definitions for materials that cause toxic effects are not always consistently used in the literature. The most common terms are toxicant, toxin, poison, toxic agent, toxic substance, and toxic chemical.

Toxicant, toxin, and poison are often used interchangeably in the literature; however, there are subtle differences as indicated below:

- A toxic agent is anything that can produce an adverse biological effect. It may be chemical, physical, or biological in form. For example, toxic agents may be chemical (such as cyanide), physical (such as radiation) and biological (such as snake venom).
- A distinction is made for diseases due to biological organisms. Those organisms that invade and multiply within the organism and produce their effects by biological activity are not classified as toxic agents. An example of this is a virus that damages cell membranes resulting in cell death.

If the invading organisms excrete chemicals which is the basis for toxicity, the excreted substances are known as biological toxins. The organisms in this case are referred to as toxic organisms. An example is tetanus. Tetanus is caused by a bacterium, Clostridium tetani. The bacteria C. tetani itself does not cause disease by invading and destroying cells. Rather, it is a toxin that is excreted by the bacteria that travels to the nervous system (a neurotoxin) that produces the disease.

A toxic substance is simply a material which has toxic properties. It may be a discrete toxic chemical or a mixture of toxic chemicals. For example, lead chromate, asbestos, and gasoline are all toxic substances. Lead chromate is a discrete toxic chemical. Asbestos is a toxic material that does not consist of an exact chemical composition but a variety of fibers and minerals. Gasoline is also a toxic substance rather than a toxic chemical in that it contains a mixture of many chemicals. Toxic substances may not always have a constant composition. For example, the composition of gasoline varies with octane level, manufacturer, time of season, etc.

- Toxic substances may be organic or inorganic in composition.
- Toxic substances may be systemic toxins or organ toxins.

A systemic toxin is one that affects the entire body or many organs rather than a specific site. For example, potassium cyanide is a systemic toxicant in that it affects virtually every cell and organ in the body by interfering with the cell's ability to utilize oxygen.

Toxicants may also affect only specific tissues or organs while not producing damage to the body as a whole. These specific sites are known as the target organs or target tissues.

- Benzene is a specific organ toxin in that it is primarily toxic to the blood-forming tissues.
- Lead is also a specific organ toxin; however, it has three target organs (central nervous system, kidney, and hematopoietic system).

A toxicant may affect a specific type of tissue (such as connective tissue) that is present in several organs. The toxic site is then referred to as the target tissue.

There are many types of body cells and they can be classified in several ways:

- Basic structure (e.g., cuboidal cells)
- Tissue type (e.g., hepatocytes of the liver)
- Germinal cells (e.g., ova and sperm)
- Somatic cells (e.g., non-reproductive cells of the body)

Germ cells are those cells that are involved in the reproductive process and can give rise to a new organism. They have only a single set of chromosomes peculiar to a specific sex. Male germ cells give rise to sperm and female germ cells develop into ova. Toxicity to germ cells can cause effects on the developing fetus (such as birth defects, abortions).

Somatic cells are all body cells except the reproductive germ cells. They have two sets (or pairs) of chromosomes. Toxicity to somatic cells causes a variety of toxic effects to the exposed individual (such as dermatitis, death, and cancer).

TRADITIONAL CHRONIC TOXICITY

The basic principles that govern the analysis of acute toxic effects are by and large directly applicable to cases of chronic toxicity. Chronic toxic analyses tend to differ from analyses of acute toxic effects primarily in that considerable emphasis must be placed on the "slow step" of the process, which causes the effect to be chronic rather than acute. This slow step is generally either a long-term accumulation of a toxic agent that is poorly excreted under ordinary conditions (e.g., lead, mercury) or an accumulation of some slowly reversed residual effect (e.g., acrylamide).

With lead as an example, the following statements can be made: Appreciable information is available on the pharmacokinetics of lead absorption, transport, storage, and excretion (Barry, 1975; Bernard, 1977; Campbell et al., 1984; Chamberlain, 1985; Marcus, 1985a-c; Rabinowitz et al., 1976); and much better information could be obtained with the aid of natural experiments such as strikes among lead-exposed workers, which can given information on the rate of decrease of blood lead levels after a reduction in exposure (Hattis, 1981).

Inhibition of heme synthesis enzymes at essentially all dose levels is well characterized (Haeger-Aronsen et al., 1974), and the inhibition of heme synthesis may be important in producing some of the neurotoxic effects of lead (Silbergeld et al., 1982), although the short- and long-term functional significance of different degrees of inhibition in different individuals is far from clear.

Effects on some measures of neurological function and kidney function are susceptible to study in reasonably straightforward ways. Effects on higher-order development of central nervous system functions are more difficult to determine because of an ignorance of basic mechanisms; however, some good studies have become available in recent years (Baker et al., 1983; Bellinger et al., 1987, 1990; HHS, 1988; Needleman et al., 1979, 1990; Waternaux et al., 1989). The impairment of very complex neurological functions by lead raises a significant issue in the application of the traditional toxicological paradigm to risk analysis. As indicated above, the usual assumption is that there is some functional reserve capacity in "normal" individuals that maintains "adequate" performance despite a "small" degree of perturbation of a biological parameter by a "low" dose of toxic material. However, if the function is already taxed to its limit in certain situations, even in the absence of exposure (perhaps for a first grader learning to read or for a developing fetus mobilizing all its available metabolic energy to grow and differentiate), and if the biological parameter being perturbed is limiting to the performance of that function, then any level of exposure may produce at least some reduction in performance. Addressing the issues of the population distribution of different functional reserve capacities, and the relationship of functional reserve capacities to specific biochemical parameters, is essential to the future research needs of risk assessment for classical chronic toxic agents. Also in the area of neurotoxicology, Silbergeld (1982) has written of the potential of new radioimmunoassay and functional measurement techniques to help shift the focus of research away from traditional morphological criteria of neurological damage toward more sensitive and sophisticated measures of performance.

Molecular Biological (Stochastic Process) Diseases

In addition to most carcinogenesis, molecular biological diseases include mutagenesis and at least some teratogenesis. The subject of quantitative risk assessment for carcinogenic hazards has been discussed extensively elsewhere (American Industrial Health Council, 1987; Bishop, 1987; Crump et al., 1976, 1977; EPA, 1986a; Hattis, 1982; Hattis and Smith, 1986; Moolgavkar, 1986; Rai and Ryzin, 1981; Whittemore, 1980). However, it is worth briefly recapitulating some basic features of the carcinogenic process and their implications for cancer dose-response relationships.

Science is now much closer than it was a decade ago to understanding the fundamental mechanisms involved in carcinogenic transformation. For some time it has been clear that tumors arise as a result of a series of changes or rearrangements of information coded in DNA within single cells (Cleaver and Bootsma, 1975; Fialkow, 1977; Hattis, 1982; Knudson, 1973, 1977; McCann et al., 1975; Vogel and Motulsky, 1979). These changes are often induced by electrophilic metabolites of the parent compounds to which organisms are exposed (Miller and Miller, 1981). With the identification of "oncogenes," some detailed molecular characterization is being provided of the changes resulting in DNA (Fischinger and DeVita, 1984; Hoel, 1985; Modali and Yang, 1984; Yunis, 1983).

It has also been apparent for some time that further headway cannot be made in elucidating the shapes of carcinogenesis dose-response relationships at low dosages simply by increasing the numbers of animals studied in conventional bioassays. A variety of mathematical models with dramatically different consequences for low-dose risk can always be found that fit the observations about equally well (Maugh, 1978; Whittemore, 1983). Low-dose risk projections are, therefore, inevitably much more determined by the choice of model than by the available data (Guess et al., 1977; Whittemore, 1980), if what is meant by "data" is restricted to observations of the incidence of ultimate adverse effects in small groups of animals.

Because of sample size limitations, animal carcinogenesis bioassays must be done within a limited range of relatively high dose levels. Typically, the difference between the minimum detectable response and a response that effectively saturates the system or causes interference through overt toxicity is only one to two orders of magnitude (often even less). Over this high dose range near levels where the agent produces overt toxic effects, enzyme saturation and other forms of pharmacokinetic nonlinearities are most likely. If in dose-response modeling for risk assessment, the nonlinearities of pharmacokinetic origin are not separated from the nonlinearities that may arise from the multiple mutation mechanism that is central to carcinogenesis, our ordinary curve-fitting procedures will implicitly attribute the pharmacokinetic nonlinearities to the fundamental carcinogenic process (Hoel et al., 1983). The resulting errors are particularly serious if one wishes to produce the best point estimates of carcinogenic risk in addition to upper confidence limits.

Clearly, to make real progress in modeling carcinogenic risks, knowledge of the fundamental processes involved must be used to break open the black box between external exposure levels and ultimate production of tumors. The use of pharmacokinetic models and intermediate parameters ("markers") to characterize the dose-response characteristics of small segments of the causal pathway to carcinogenesis has considerable potential to improve dose-response modeling for the process as a whole (Hattis, 1988). Such markers can include both those that may lie directly along the causal pathway, such as DNA adduct formation, and putative correlates, such as hemoglobin adduct formation, that can be good indicators of the concentration-time product of active intermediates in the systemic circulation.

One key fact must be recognized from the beginning about pharmacokinetic modeling, however. Whatever nonlinearities may be produced at high doses by the saturation of enzymes, the saturation of active transport processes, the depletion of cellular reactants for electrophilic agents, or changes in cell division rates to make up for cell killing due to overt toxicity, all of these nonlinearities must necessarily disappear as one approaches very low dose rates (Hattis, 1990a). The slope of the line relating ultimate DNA lesions in replicating cells to external dose may well be very different at low than at high doses, but it must be linear. The basic reason for this is that at low doses the rates of the transport and transformation processes that lead to DNA damage and repair depend directly on the number of collisions between molecules of an "input" chemical (or activated intermediate or DNA adduct) and a resident cellular reactant (or hole in a membrane or repair enzyme molecule). At low doses the number of resident cellular reactant molecules does not change appreciably as a function of the concentration of the input. Therefore, the number of relevant collisions and the rates of reactions and side reactions in the causal sequence at low dosage must be direct linear functions of the amounts of input chemical and its activated derivatives. Some finite fraction of the ultimate DNA lesions must escape repair before the next cell replication as long as the cells affected have a nonzero turnover rate, there are a finite number of repair enzyme molecules, and the repair molecules operate at a finite rate.

CRITIQUE OF RISK ASSESSMENTS

One of the charges to the committee was to review and critique previous agency efforts to assess the likely human risks of chemical contaminants of aquatic organisms. The committee was asked to suggest directions for the improvement of risk assessment approaches to better serve the needs of decision making on chemical risks, but not to attempt to develop its own set of final risk numbers. The committee has chosen to focus on PCBs and methylmercury as two paradigmatic cases, representing one carcinogen and one noncarcinogen, for which there is a relatively sufficient public record of FDA efforts to assess the risks and benefits of alternative control options.

The PCBs are of special interest in light of the results presented in this chapter. By using conventional carcinogenic risk assessment techniques (although these have considerable difficulties in general and especially as applied to PCBs) and available data on likely average daily intakes in the United States, PCBs are shown to pose by far the largest *potential* carcinogenic risk of any environmental contaminants for which measurements exist. According to EPA's cancer potency factor, the aggregate U.S. lifetime risk could be as high as 6×10^{-5}, corresponding to approximately 120 cancers per year. With FDA's lower estimate of the upper confidence limit (UCL) of the PCB cancer potency, the aggregate risk for the U.S. population would be considerably less, but still not insignificant, 2.7×10^{-6}, implying a maximum toll from commercial seafood of about 5 cancers per year.

These risks would be much greater for the subpopulation consuming relatively large amounts of sport or subsistence caught fish near the FDA tolerance level of 2 parts per million (ppm). For someone who consumes 20 kilograms (kg) per year of 20-ppm fish [which is probably not far from the actual consumption of some izndividuals in some areas of Lake Michigan

(Humphrey, 1983a,b)], the upper-confidence-limit lifetime risk that would be expected by using the FDA cancer potency estimates is about 5×10^{-4}; for the EPA cancer potency estimate, the lifetime upper-confidence-limit risk would be somewhat more than 1%. Perhaps in part because of the difference in cancer potency factors, there is also a great difference in the practical policies of the two agencies in regulating PCBs, with FDA generally taking a far less protective posture toward PCB exposure than EPA.

Methylmercury is also of special interest because it shows the operation of the old no-effect level/safety factor paradigm in a practical case where, as it happens, an alternative type of quantitative assessment is also possible based on existing data. In this case, therefore, one can examine what levels of risk might implicitly be accepted by using FDA's rule-of-thumb procedure for dealing with noncancer effects.

In the following material, the committee will assess for both cases the reasonableness of the calculations that went into FDA's decision making in terms of the information available at the time; and
the improvements that could and should be made by applying more modern analytical techniques to recently available or feasibly obtainable data on population exposures, the nature and mechanisms of potential adverse effects, and likely population dose-response relationships.

To briefly foreshadow the committee's conclusions, three statements can be made:

1. Both assessments (Cordle, 1983; Cordle et al., 1982; FDA, 1979; Tollefson and Cordle, 1986), even at the times they were done, suffered from ad hoc unsystematic approaches to the treatment of population heterogeneity in dosage and (for methylmercury) susceptibility to adverse effects. For mercury, the likely special susceptibility of developing fetuses was mentioned in discussion. However, a tenfold "safety factor" was applied to the lowest blood level reported to produce effects for adults (rather than to a no-effect level of intake, which would have been more consistent with established procedures) in the cited epidemiological studies, without numerical allowance for extra sensitivity of fetuses, without a quantitative dose-response treatment of the data then available for adults to gauge the potential adult risk at the blood level selected as the highest permissible for U.S. consumers, and without a quantitative treatment of the effects of pharmacokinetic differences among people that would tend to make blood levels and risks as a function of dietary intake more variable than blood levels themselves.
2. Procedures for the systematic quantitative analysis and communication of uncertainties are absent. Both analyses use the "method" of

compounding a series of ad hoc allegedly "conservative" assumptions (such as the use of the consumption level of the 90th percentile consumer, the upper 99% confidence limit for the PCB animal potency calculation, but a body weight rather than a surface area scaling rule to translate animal to human dose). The difficulty in picking a single series of point estimates of uncertain parameter values is that after the first few such assumptions have been entered into an analysis, no person on earth can determine where one actually is on an overall distribution of the likelihoods of different outcomes. Monte Carlo simulation procedures are now readily available via personal computer-based software for the calculation of probability density functions of different outcomes, given uncertainties in multiple parameters affecting population risk. Such procedures have been effectively advanced in the context of risk assessment in a recent report by Finkel (1990).

3. Both assessments have been rendered substantially obsolete by the development of recent information related to PCB and to methylmercury risks, changes in the economic conditions assumed in the cost analyses, and possibly changes in residue levels and available options for population exposure reduction.

In the case of PCBs, understanding of mechanisms has advanced to the point where there are now serious proposals for a congener-specific assessment of relative activity (Clement Associates, 1989; Jones, 1988). In the view of the committee, this congener-specific activity analysis must be coupled with a congener-specific treatment of pharmacokinetics, which now seems possible based in part on the human dietary exposure and serum concentration data being collected by Humphrey and coworkers (H. Humphrey, Michigan Department of Public Health, Lansing, personal communication, 1989). Both congener-specific activity estimates and pharmacokinetics should be used to evaluate not only existing animal carcinogenesis data, but also recently emerging epidemiological information from studies in workers.

In the case of methylmercury, recent data allow preliminary evaluation of the population dose-response relationship for a number of long-term neurological effects from exposure during fetal development (Marsh et al., 1987). The analysis in the following section indicates that the apparent interindividual variability in susceptibility for fetal effects is much broader than that for adult effects. Consequently, although the tenfold safety factor, as applied, appears to offer a reasonable degree of protection for adult effects, projections based on an additional toxicological assumption of lognormal distribution of threshold for effects (logprobit projections) of the fetal dose-response data suggest the possibility of appreciable risk from methylmercury exposure, even at levels to which many people are exposed via the diet.

Published projections applying a logistic model to the same data come to similar conclusions (Cox et al., 1989).

TOXIC EFFECTS

Toxicity is complex with many influencing factors; dosage is the most important. Xenobiotics cause many types of toxicity by a variety of mechanisms. Some chemicals are themselves toxic. Others must be metabolized (chemically changed within the body) before they cause toxicity.

Many xenobiotics distribute in the body and often affect only specific target organs. Others, however, can damage any cell or tissue that they contact. The target organs that are affected may vary depending on dosage and route of exposure. For example, the target for a chemical after acute exposure may be the nervous system, but after chronic exposure the liver.

Toxicity can result from adverse cellular, biochemical, or macromolecular changes.

Examples are:

- Cell replacement
- Damage to an enzyme system
- Disruption of protein synthesis
- Production of reactive chemicals in cells
- DNA damage
- Some xenobiotics may also act indirectly by:
- Modification of an essential biochemical function
- Interference with nutrition
- Alteration of a physiological mechanism
- Factors Influencing Toxicity
- The toxicity of a substance depends on the following:
- Form and innate chemical activity
- Dosage, especially dose-time relationship
- Exposure route
- Species
- Age
- Sex
- Ability to be absorbed
- Metabolism
- Distribution within the body
- Excretion
- Presence of other chemicals.

The form of a substance may have a profound impact on its toxicity especially for metallic elements. For example, the toxicity of mercury vapor differs greatly from methyl mercury. Another example is chromium. Cr^{3+} is relatively nontoxic whereas Cr^{6+} causes skin or nasal corrosion and lung cancer. The innate chemical activity of substances also varies greatly. Some can quickly damage cells causing immediate cell death. Others slowly interfere only with a cell's function.

For example: hydrogen cyanide binds to cytochrome oxidase resulting in cellular hypoxia and rapid death, nicotine binds to cholinergic receptors in the CNS altering nerve conduction and inducing gradual onset of paralysis. Exposure route is important in determining toxicity. Some chemicals may be highly toxic by one route but not by others.

Two major reasons are differences in absorption and distribution within the body. For example: ingested chemicals, when absorbed from the intestine, distribute first to the liver and may be immediately detoxified, inhaled toxicants immediately enter the general blood circulation and can distribute throughout the body prior to being detoxified by the liver.

Frequently there are different target organs for different routes of exposure.

Toxic responses can vary substantially depending on the species. Most species differences are attributable to differences in metabolism. Others may be due to anatomical or physiological differences. For example, rats cannot vomit and expel toxicants before they are absorbed or cause severe irritation, whereas humans and dogs are capable of vomiting. Selective toxicity refers to species differences in toxicity between two species simultaneously exposed. This is the basis for the effectiveness of pesticides and drugs.

Examples are: an insecticide is lethal to insects but relatively nontoxic to animals, antibiotics are selectively toxic to microorganisms while virtually nontoxic to humans.

Age may be important in determining the response to toxicants. Some chemicals are more toxic to infants or the elderly than to young adults.

For example: parathion is more toxic to young animals, nitrosamines are more carcinogenic to newborn or young animals. Although uncommon, toxic responses can vary depending on sex.

Examples are: male rats are 10 times more sensitive than females to liver damage from DDT, female rats are twice as sensitive to parathion as male rats.

The ability to be absorbed is essential for systemic toxicity to occur. Some chemicals are readily absorbed and others poorly absorbed. Nearly all alcohols are readily absorbed when ingested, whereas there is virtually no absorption for most polymers. The rates and extent of absorption may vary greatly depending on the form of the chemical and the route of exposure. For example: ethanol is readily absorbed from the gastrointestinal tract but poorly absorbed through the skin, organic mercury is readily absorbed from the gastrointestinal tract; inorganic lead sulfate is not.

Metabolism, also known as biotransformation, is a major factor in determining toxicity. The products of metabolism are known as metabolites. There are two types of metabolism - detoxification and bioactivation. Detoxification is the process by which a xenobiotic is converted to a less toxic form. This is a natural defense mechanism of the organism. Generally the detoxification process converts lipid-soluble compounds to polar compounds. Bioactivation is the process by which a xenobiotic may be converted to more reactive or toxic forms.

The distribution of toxicants and toxic metabolites throughout the body ultimately determines the sites where toxicity occurs. A major determinant of whether or not a toxicant will damage cells is its lipid solubility. If a toxicant is lipid-soluble it readily penetrates cell membranes. Many toxicants are stored in the body. Fat tissue, liver, kidney, and bone are the most common storage depots. Blood serves as the main avenue for distribution. Lymph also distributes some materials.

The site and rate of excretion is another major factor affecting the toxicity of a xenobiotic. The kidney is the primary excretory organ, followed by the gastrointestinal tract, and the lungs (for gases). Xenobiotics may also be excreted in sweat, tears, and milk.

A large volume of blood serum is filtered through the kidney. Lipid-soluble toxicants are reabsorbed and concentrated in kidney cells. Impaired kidney function causes slower elimination of toxicants and increases their toxic potential.

The presence of other chemicals may decrease toxicity (antagonism), add to toxicity (additivity), or increase toxicity (synergism or potentiation) of some xenobiotics.

TOXICOLOGY CONSIDERED : AN ART AND SCIENCE

Centuries ago, Paraceulsus (1493-1541) one of the founding fathers of toxicology, noted, "All things are poisons, for there is nothing without poisonous qualities. It is only the dose which makes a thing a poison." Therefore, toxicology is defined as the study of the adverse (toxic) effects of both chemical and physical agents on biological systems.

Toxicology is considered both an art and science; the science involves the data-gathering phase while the art of toxicology uses this acquired data to predict the likelihood of adverse responses in humans and wildlife following exposure to one or more agents in situations where there is limited information.

For example, the observation that exposure to chloroform can produce hepatomas in mice is a documented fact while the conclusion that it will do so in man is a prediction or hypothesis.

There are two classes of agents that toxicologists study:

- Chemical agents (synthetic or naturally-occurring) that include 64,000 compounds used in commerce (5,000 million tons per year or more); 5,000 food additives; 4,000 medicinal drugs; 1,200 household products (at least!); substances produced by bacteria, fungi, molds, spores, plants and animals and approximately 700 new chemicals that are identified or introduced each year.
- Physical agents such as radiation (e.g. UV, electromagnetic), noise, pressure and dust.

Consequently, toxicology is an integral component of risk assessment, which uses the data obtained from toxicological studies to establish regulatory policy regarding exposure to potentially hazardous substances. In order to accurately predict the potential risk of a substance, a toxicologist requires expertise in a number of disciplines including biology, chemistry, physics, immunology, pathology, physiology, molecular biology, developmental biology, public health, epidemiology and risk assessment.

In addition to safety assessments, research into the mechanism of action of toxic substances has provided new insights into important physiological functions resulting in a better understanding of disease processes as well as to the development of new therapeutic strategies.

Fields of Study in Toxicology

It is difficult classify a toxicologist with any one particular field of toxicology since many of the problems require an integrative approach to problem solving. Therefore, most toxicologists use a variety of approaches that span several of these areas to obtain a comprehensive picture of the risks associated with the substance under investigation. These fields include:

Biochemical Toxicology

Biochemical toxicology uses the methods of biochemistry, cell biology and cell physiology to determine how toxic substances adversely affect the normal functioning of cellular components. Biochemical toxicologists investigate the enzymes responsible for the metabolism of a specific compound or how a substance inhibits an enzyme activity, which is essential for normal cell function.

Molecular Toxicology

Molecular toxicology involves the use of molecular biology to investigate the effect of toxic substances at the level of DNA. Molecular biology techniques have been used to examine the effects on compounds on gene regulation and DNA mutagenesis. These methodologies have also been used to investigate the role of proteins in drug metabolism and to identify specific domains, which are essential enzyme activity.

Chemical Carcinogenesis

Chemical carcinogenesis investigates how compounds initiate and promote the development of cancer. These studies include identifying the enzymes responsible for activating the compound to a carcinogen as well as the genes that are mutated by the compound. A few compounds that have been investigated include chemotherapeutic drugs, polycyclic aromatic hydrocarbons, aromatic amines, halogenated hydrocarbons and mycotoxins.

Environmental Toxicology

Environmental toxicology examines the effect of substances that are present in the environment. This includes investigating the effects of all types of pollution resulting from compounds that have been released into the environment following their use or that have been introduced as unwanted by-products or decomposition products. In addition to examining the effects of these compounds on mammalian species, environmental toxicologists also examine effect of these substances on birds, fish, reptiles and plants.

Genetic Toxicology

Genetic toxicology is the study of the interaction of chemical and physical agents with the process of heredity. These studies involve investigating the type of mutations that an agent causes and its effects on chromosomes. In addition, genetic toxicologists develop assays to identify genotoxic substances that may adversely affect human health and environmental quality.

Developmental Toxicology

Developmental is concerned with the investigation of chemically induced teratogenic effects or birth defects. Studies involve examining any detrimental effect produced following exposure of a developing organism during development. Currently, developmental biology is one of the most exciting fields of study which will facilitate a better understanding of how toxic substances perturb the complex regulation patterns required for normal development.

Immunotoxicology

The field of immunotoxicology examines the effect of substances on the immune system. Studies involve examining the ability of chemicals to compromise immune function, elicit hypersensitive reactions and induce autoimmunity.

Reproductive Toxicology

The survival of any species is dependent on the integrity of its reproductive system. Therefore, the effects of drugs and environmental chemicals on reproductive fitness are a major health concern. Reproductive toxicologists examine how toxic substances affect reproductive capacity and develop assays to identify agents that may adversely affect reproductive performance.

In addition to these fields of study, there are several other areas of toxicology that have not been listed including regulatory toxicology (risk assessment/risk management), forensic toxicology, clinical toxicology, veterinary toxicology and occupational toxicology.

Effect of Toxicology

Toxicology is the study of the adverse effects of chemicals on living systems, whether they be human, animal, plant or microbe. 'Adverse effect' can range from a life threatening injury to something that might be considered a minor annoyance. A mild irritant that causes watering of the eyes may not seem like a significant or real adverse effect, but for the person operating fine equipment, such effect can be not only deleterious but potentially dangerous.

Toxicology is an inter-disciplinary science that integrates the principles and methods of many fields: chemistry, biology, pharmacology, molecular biology, physiology and medicine. For example, by combining the study of the physiological effects of certain chemical structures and the molecular biological mechanisms that explain those effects, the toxicology can provide better understanding of the actions of a class of chemical substances, a finding of value to the chemist.

In the practice of toxicology other disciplines are also frequently relied upon. Hyrdogeologists, environmental engineers and air modelers are often used to determine the route, concentration and migration of chemicals in water, soil and air, respectively. Clinicians are relied on to record signs and symptoms and conduct differential diagnoses, to determine if a certain disease process is present and to eliminate other, non environmental causes.

The scientific pursuit of toxicology is typically divided between *observational studies*—looking at what effects result from exposure to a particular substance—and *mechanistic studies,* which attempt to understand and explain the basis for such effects. These two activities form the basis of toxicology as an experimental science, that is, that which takes place in the lab or occasionally in the field, as one usually thinks of science.

Risk assessment is another branch of toxicology which attempts to determine, probabilistically, outside the laboratory, the likelihood of an adverse effect based on a particular exposure scenario. This is not experimental activity, involves much more uncertainty about its findings and is as much art as science. It is conducted quantitatively, relying on mathematics and computer modeling, or qualitatively, relying more on experience and similar scenarios that have been previously looked at.

Regulatory toxicology relies on risk assessment and experimental data to determine the risk and benefits, or the costs and benefits of exposure to certain chemicals, to determine whether such chemical will be allowed in the public sphere and to what extent its use and exposure will be regulated. This field of toxicology probably has the greatest effect on our daily lives of all the different fields of toxicology.

Clinical toxicology involves the application of toxicological principles within a diagnostic setting, usually to determine whether a presenting adverse effect or disease or injury is due to some type of chemical exposure. This area of toxicology is typically practiced by a physician, nurse or other clinician, often times in consultation with the experimental toxicologist, who is in a position to better explain certain published experimental findings and whether they would be applicable to the case at hand.

Forensic toxicology focuses on medical-legal aspects of chemical exposure and toxic injury. Evaluation of analytical chemistry techniques and measurements, and testimony of conclusions and opinions in courts of law encompass the activity of this specialized field of toxicology.

The concept of dose is a basic principle in toxicology. The phrase, 'dose makes the poison' does capture a fundamental principle of toxicology, but the actual study of toxicology is much more nuanced and complex than this simple statement suggests.

Dose is the amount of chemical that comes into contact with the body or gets inside the body. Exposure is the amount of chemical in a media—air, water, soil or food—that is directly available to the body (through breathing, drinking, touching or consuming). Exposure includes both the concentration of the chemical in the media and the length of time such chemical is available to the body (concentration x time).

Although exposure is important in determining dose, other factors also come into play. For example, if two individuals are exposed to the same amount of chemical in the air—same concentration and for same length of time—one will inhale a much higher dose than the other if that person is exercising more vigorously. Even though exposure is identical for the two individuals, the faster and deeper breathing of the first individual will pull more chemical (i.e., dose) into the lungs.

Also, not all doses are equivalent in their toxicity even if they are equivalent in size. A high concentration of a chemical exposed for a short time can produce the same dose as a low concentration exposed for an extended period of time, but the effect of these equivalent doses may be quite different because of the concentrations of the chemical.

An acid, for example, requires a minimum concentration to exert a corrosive effect (that is, permanently damage tissue); below that threshold an acidic effect will not occur no matter how long exposure takes place (i.e., how large the total dose). So even though a low concentration of chemical for an extended period of time can produce the same dose as a high concentration for a short period of time, such effect will not be acidic (corrosive).

Dose does make the poison but so do other factors, such as concentration of the chemical and the threshold required to exert a particular effect. Important in the study of toxicology is understanding these different factors and how they intereact, and then determining which are of primary importance in each exposure scenario.

TYPES OF POISONING SUBSTANCES

It has been customary to subdivide poisoning into acute, a sudden violent syndrome caused by a single large dose of poison and chronic, a persistent, lingering condition brought on by small repeated doses, with subacute poisoning somewhere in between. This subdivision, however, is not really tenable, as there are types of poisoning which would be difficult to fit into any of these categories. 'Chronic'copper poisoning in sheep only becomes manifesting an acute hemolytic crisis. Symptoms of bracken poisoning may not appear until months after the plant has been ingested. In addition to these types of poisoning, there are other unto- ward effects due to 'poisonous' substances.

- Allergy, an immunological response due to sensitization of the subject by a previous dose, although less common in animals than in man,is still quite well known.
- Carcinogenicity, in which the agent is responsible for the formation of neoplasia, bracken and the cycads are examples.
- Teratogenicity, in which material ingested by the mother at some definite stage in pregnancy produces abnormalities in the off- spring. The classical example in veterinary toxicology is the cyclopean malformation in lambs due to Veratrum californicum.

Metabolism of Toxicants

Absorption

In order to exert their toxic effects most poisons must be absorbed into the blood stream.

Under natural conditions toxicants may enter the body through the lungs, gut and skin. Substances maybe injected into the body subcutaneously, intramuscularly, intravenously and intraperitoneally.

- *Respiratory tract*: The very extensive highly vascular pulmonary mucous membrane affords an excellent channel of absorption for gases and for solids and liquids, particularly when in a fine state of dispersal (as aerosols or dusts).
- *Alimentary tract*: It is the most usual route of entry of a poison. In all species, much absorption occurs in the small intestine; in the dog it can also take place from the stomach, in the ruminant from the rumen and reticulum, and in all species, especially in non-ruminant herbivora, from the large gut. The contents of the alimentary tract can modify the action of poison.

  The hydrochloric acid present in gastric juice may aid the solution of originally insoluble materials. A full stomach or rumen may delay symptoms of poisoning or may dilute the poison to such an extent that it is relatively harmless.
- *Dermal exposure*: The unbroken skin does not offer a favorable channel of absorption to most compounds. Nicotine is one of the exceptions which in aqueous solutions,is very efficiently absorbed through intact skin. Absorption takes place more readily from oily solutions or emulsions.

Absorption through damaged or abraded skin or a wound occurs as through moist mucous membranes. Subcutaneous or intramuscular injection is equally effective, while intravenous administration is the fastest way of introducing a poison into the blood stream.

Distribution and Accumulation

Hepatic Storage

All foreign compounds entering the body pass to the liver which is the major detoxifying organ in the body and by virtue of this fact many poisons accumulate in this organ. It is therefore not unexpected to find hepatic lesions as a consequence of exposure to many toxicants.

Extra Hepatic Storage

Some poisons are selectively deposited in certain organs or tissues. Iodine is largely taken up by the thyroid glands; strontium, fluorine and lead are deposited in the bones. A knowledge of the distribution of a particular poison is of great help in the selection of organs for chemical analysis.

Biotransformation and Detoxication

Many foreign compounds that are introduced into the body undergo chemical transformation, and this process is generally referred to as metabolic transformation or biotransformation. Very frequently this process is referred to as the detoxication mechanism. However, this term is misleading because the metabolic transformation of a foreign compound may result into increased or decreased toxicity of the product. Two categories of enzyme systems are known to exist in mammals that cause chemical transformation of foreign compounds.

One category consists of enzymes that normally occur in the tissues and are responsible for transformation of normal endogenous substances in tissues as well as foreign chemical e.g., alcohol dehydrogenase. Second category consists of an enzymes system that is very important in toxicology. These are known as drug metabolizing enzymes, which are present in many tissues but are particularly in liver cells. These enzymes are located in small particles called microsomes which in turn are located in the smooth surface of endoplasmic reticulum. These microsomal enzymes are capable of catalyzing a variety of biotransformation reactions, the major ones being oxidation, reduction, and hydrolysis.

The drug metabolizing enzymes system is also called as mixed function oxidase (MFO) enzyme system for the obvious reason that is capable of catalyzing various types of reactions. The oxidase system requires the presence of a cofactor NADPH, and molecular oxygen for its activity. NADPH reduces a component of the microsomes which reacts with molecular oxygen to form an active oxygen intermediate which oxidizes the drug or the foreign chemical. The components of microsomes which is reduced by NADPH is a heme protein called cytochrome P-450.

R.T. Williams divided the biotransformation mechanism of foreign chemicals into two major types:

- The non synthetic reactions involving oxidation, reduction and hydrolysis,
- the synthetic or conjugations involving production of a product that is biosynthesized from the chemical (or its metabolite) plus an endogenous metabolite such as glycine.

If we call the non synthetic reactions as phase 1 of biotransformation mechanism, then conjugation reactions will be phase II, which are described below.

- Conjugation Reactions: These are reactions in which a poisoner ametabolite is combined with some compound provided by the animal body. Following is the list of some of these processes.

- Glycine Conjugations: Benzoic acid combines with glycine to form hippuric acid.
- Glucuronic Acid: Conjugation with this occurs in may cases. e.g.,dinitrophenol, chloral hydrate.
- Sulfate Conjugation: Some phenols conjugate with $=SO_4$ to form ethereal sulfates.
- Cysteine Conjugation: Arsenic trioxide and mainly benzene, polycyclic hydrocarbons, and certain halogenated hydrocarbons conjugate with cysteine to form mercapturic acids.
- Acetylation Reactions: This takes place between CoASH and amino groups of aromatic compounds, e.g., sulfanilamide derivatives. Dog is deficient to acetylate aromatic amine groups.
- Methylation Reactions: It is relatively uncommon form of detoxication and is confined to the heterocyclic nitrogen atoms in compound of the pyridine and quinoline type.
- Thiocyanate Formations: Inorganic cyanides are converted by the enzymerhodanese to thiocyanate. It is a true detoxication, for sodium thiocyanate is nearly 200 times less toxic than sodium cyanide and is slowly excreted in the urine.
- Glutamine and Ornithine Conjugation: Conjugation processes with glutamine and ornithine occur in man and bird, respectively.

Inhibition of Biotransformation Mechanisms

The microsomal enzyme systems can be inhibited by several compounds in concentrations which by themselves have little pharmacologic activity. The most common example of such inhibitor is SKF 525A (diethyl aminoethanol ester of diphenyl propyl acetic acid).

Thus, the inhibition of the microsomal enzyme system would result into the inhibition of the metabolism of toxic foreign compound, the presence of which may cause increased toxicity.

On the other hand, when the product of metabolic transformation is of greater toxicity than that of the parent compound, The inhibition of the microsomal enzymes by SKF525A would be expected to protect the animal from toxicity resulting from metabolism of the parent compound.

Induction of Biotransformation Mechanisms

The total quantity of microsomal drug metabolizing enzymes can be increased in humans and in animals by prior administration of large variety of chemical substances.

Such substances include the anesthetics, such as nitrous oxide; the sedatives, such as barbiturates; the analgesics, such as phenylbutazone; and the insecticides, such as chlordane. Induction of increased enzyme activity usually involves repeated exposure to the inducing agent. It is usually temporary and lasts for two to four weeks following the administration of the inducing chemical. Since metabolic transformation has been shown to result in the formation of more or less toxic products as compared to the parent compound, enzyme induction may be protective to the animal (when detoxication is involved) or detrimental to the animal (when toxication is involved).

Elimination

Following are the major ways through which poisons and their metabolites are excreted.

- Fecal excretion. Ingestation of a relatively insoluble poison (e.g., lead arsenate) is followed by excretion of the major part in the feces. Substances may also find their way into faeces via bile; metals stored in the liver are slowly excreted in this way.
- Pulmonary excretion. Volatile poisons may be mainly excreted in the expired air, e.g., CS2, cyanide. In phosphorus poisoning the breath may smell of garlic odor and glow in the dark. Diagnosis of hemlock poisoning may be made from the characteristic "mouse-like" odor of coniine in the exhaled air (and also in urine). The lesions found in the lungs in paraffin poisoning are probably due to irritant effect caused by pulmonary excretion.
- Urinary excretion. This is the most important pathway of the excretion of a poison. Irritant poisons cause damage to kidney. Urine is often a convenient material for diagnostic analysis. In veterinary field it is of great importance in detecting the pasture contamination with fluorides.
- Milk and dermal excretion. Excretion can also take place through skin, e.g., arsenic, and in lactating animals in milk. Many of the insecticides are fat soluble and it has been shown that DDT, aldrin, and several other chlorinated hydrocarbons can be detected in minute amounts in cow's milk.

Toxicokinetics

Toxicokinetics involves the relationship between tissue concentration of a toxicant and time. There are two types of kinetics:

1. Zero-order kinetics
2. First order kinetics.

In zero-order kinetics, a toxicant is eliminated from the body at a fixed amount per unit time. This occurs when the toxicant elimination process is saturated and a fixed amount of toxicant (e.g., mg or g) is excreted per unit time.

Many toxicants are eliminated from animal tissues in a fixed-order manner. Thus, in first-order kinetics a constant fraction of a toxicant/drug is eliminated per unit time. For example, if the initial level were 500 mg and it took ten days to reach 250 mg, then it will take additional 10 days to reach 125 mg. During the first 10 days a total of 250 mg toxicant was excreted and during the next 10 days only 125 mg. In each case, the amount excreted was a constant fraction (i.e., 50 per cent)of the original amount.

The ten-day period is called half-life of the toxicant. Various species handle excretion of a toxicant in different manner and therefore a given toxicant will have different half-lives in different species.

MODERN TOXICOLOGY

Toxicology has drawn its strength and diversity from its proclivity to borrowing from almost all the basic sciences to test its hypotheses. This fact, coupled with the health and occupational regulations that have driven toxicology research since, 1900, has made this discipline exceptional in the history of science. With the advent of anesthetics and disinfectants in the late 1850s, toxicology as it is currently understood began.

The prevalent use of "patent" medicines led to several incidents of poisonings from these medicaments, which, when coupled with the response to Upton Sinclair's exposé of the meatpacking industry in *The Jungle,* culminated in the passage of the Wiley Bill in 1906, the first of many U.S., pure food and drug laws. During the 1890s and early 1900s, the discovery of radioactivity and the vitamins, or "vital amines," led to the use of the first large-scale bioassays (multiple animal studies) to determine whether these "new" chemicals were beneficial or harmful to laboratory animals. One of the first journals expressly dedicated to experimental toxicology, *Archiv für Toxikologie,* began publication in Europe in 1930.

That same year the National Institutes of Health (NIH) was established in the United States. As a response to the tragic consequences of acute kidney failure after taking sulfanilamide in glycol solutions, the Copeland bill was passed in 1938. This was the second major bill involving the formation of the U.S. Food and Drug Administration (FDA). The first major U.S., pesticide act was signed into law in 1947. The significance of the initial Federal Insecticide, Fungicide, and Rodenticide Act was that for the first time in U.S., history a substance that was neither a drug nor a food had to be shown to be safe and efficacious.

AFTER WORLD WAR II

- You too can be a toxicologist in two easy lessons, each of ten years.—*Arnold Lehman* (circa 1955)

The mid-1950s witnessed the strengthening of the U.S. FDA's commitment to toxicology. The U.S. Congress passed and the president of the United States signed the additives amendments to the Food, Drug, and Cosmetic Act.

The Delaney clause (1958) of these amendments stated broadly that any chemical found to be carcinogenic in laboratory animals or humans could not be added to the U.S., food supply. Delaney became a battle cry for many groups and resulted in the inclusion at a new level of biostatisticians and mathematical modelers in the field of toxicology. Shortly after the Delaney amendment, the first American journal dedicated to toxicology, *Toxicology and Applied Pharmacology,* was launched. The founding of the Society of Toxicology followed shortly afterwards.

The 1960s started with the tragic thalidomide incident, in which several thousand children were born with serious birth defects, and the publication of Rachel Carson's *Silent Spring* (1962). Attempts to understand the effects of chemicals on the embryo and fetus and on the environment as a whole gained momentum. New legislation was passed, and new journals were founded. Cellular and molecular toxicology developed as a subdiscipline, and risk assessment became a major product of toxicologic investigations. Currently, many dozens of professional, governmental, and other scientific organisations with thousands of members and over 120 journals are dedicated to toxicology and related disciplines. In addition, the International Congress of Toxicology is made up of toxicology societies from Europe, South America, Asia, Africa, and Australia, which brings together the broadest representation of toxicologists.

4

Environmental Microorganisms in Biogeochemical Cycles

AQUATIC MICROBES: IMPACT ON MAN AND ENVIRONMENT

The aquatic environment is that part of the earth covered with water and it is alled the hydrosphere. Air is the atmosphere while soil is the lithosphere. ydrosphere, atmosphere and lithosphere comprise the biosphere, i.e. an nvironment where living things are found. Microbes are minute living hings that occur in all environments and are therefore ubiquitous. The environment has received more attention globally and locally in the last decade after the United Nations Earth Summit in Rio de Janeiro, 1992. As Professor of Environmental Microbiology, the biosphere is my research domain. But of particular interest to me is the hydrosphere, my specialty being aquatic microbial ecology and water pollution. So, I have chosen for my inaugural lecture the topic, "Aquatic microbes: impact on man and environment".

Microorganisms in the aquatic environment affect man either positively or negatively, directly or indirectly. In other words, these minute living things may be beneficial or harmful to man, his activities as well as other natural resources. For purposes of completeness, it is noted that the presence of some microorganisms in water may sometimes have no apparent or visible effect, being neutral to the ecosystem. The ecosystem is the highest organization in the hierarchy of life. It covers individual organisms in the food chain, populations, energy flow, biogeochemical cycles, interactions between living and non- living resources, as well as human impact. Microorganisms are involved in all of these activities hence man should tåke cognizance of microorganisms in sustainable development.

Discovery of Microorganisms

For a long time man was not aware of the existence of the microbial world. At that time, activities and impacts of microbes were attributed to the supernatural. The reason is not far fetched. A microbe is invisible to the unaided human eye. In other words our limit of vision does not cover the size of a microbe. We can all visualize one metre length, one centimetre i.e. one hundredth of a metre. But a smaller division of a metre such as one millionth or billionth is not discernable by the unaided human eye. Thanks to Physics and the advent of magnifying lenses. The microscope is used to enlarge cells so that the morphology or shape of a microbial cell can be discerned. It was Antony van Leewenhoeck (1677) who in 1676 made crude microscopes and, out of curiosity, used them to examine various specimens including well- and sea- water. He discovered what he called 'animacules' or 'minute animals'. van Leewenhoeck thus initiated the study of aquatic microbes and is indeed the father of Environmental Microbiology. Whereas a bacterial cell is not visible by the naked human eye, a population of bacteria may become visible after the individual cells are cultured having been supplied with nutrients for growth and multiplication to the size of a mass of cells or a colony. Even today, it is not easy to appreciate and acknowledge the presence of microorganisms. Man tends to relate to the effect, result or symptoms produced by the activities of microorganisms.

What of fermentation of palm juice into palm wine (Odeyemi & Akpata,1988)? Salmonellosis contracted from poultry (Akpata & Fashola, 1985)? Anyone would give prompt attention to gastroenteritis (upset stomach) from cholera or dysentery due to drinking contaminated water or eating molluscs harvested from a polluted bay. If microorganisms cause changes that are pleasant, the process is maintained, if undesirable or harmful, the process is terminated or controlled.

Position of Aquatic Microbes in the Evolution of Life

Let us look at microbes in water, their evolution and niches. Water as a chemical (H2O) is devoid of carbon/energy and nitrogen sources of nutrients required for the growth of living things. It is the presence of extraneous substances in water that provides the nutrients for microbial growth. These microbes affect the quality of water and thereby influence the use that man makes of it.

There are microorganisms in water. Some are indigenous to water and are called autochthonous species while some are foreign or allochthonous species that find their way into the aquatic environment. Autochthonous species are involved in the cleansing of water, utilizing the little nutrients available because a clean body of water is referred to as oligotrophic, having low nutrient content. When wastes are loaded into water it becomes eutrophic providing nutrients for microbial proliferation. In this way, eutrophication causes ecosystem dis-equilibrium, initiating signs of pollution with ultimate loss of the value of the aquatic resources. Water is an essential natural resource for sustainability of life. It is known that a living organism requires water at some stage of its life cycle: the human, the developing embryo or foetus, the mosquito larva, the germinating plant seed or microbial spore requires water for survival. Some arid lands require only irrigation for vegetation to develop!

Indeed the abundant water on the surface of planet Earth millions of years ago, was the conducive environment for the formation of molecules into the cytoplasmic broth that constituted the precursors of the first living things.

These led to the evolution of the primitive cell (protocaryote) or prokaryote in which the nucleus is naked. Is it not interesting that bacteria and bluegreen algae belong to this group? The eucaryotic cell of fungi, protozoa and indeed multicellular organisms, including man, evolved later with a nuclear membrane and more complex organisation. It is noteworthy that the biochemical activities of microorganisms contributed a lot to the maintenance of an oxygen atmosphere responsible for the existence of life on earth.

Thus, microorganisms evolved in water and to date various species remain in their habitat of origin occupying different niches in the aquatic environment. Microbes abound in surface waters: fresh, brackish and marine. They occur in rivers, lakes, estuaries and the seas. By their microscopic size and versatile metabolic capabilities, microbes find niches and can colonize the biosphere. The number of microorganisms in surface waters is high because of the input from natural, and anthropogenic (manmade) sources including various industries. During rainfall, natural run-off washes soil microorganisms into rivers and streams that become turbid.

Also, urbanization creates large populations that generate wastes e.g. refuse (solid) and sewage (liquid). The disposal of these waste materials remains a colossal problem. There is the urbanization of farm animals and general agricultural practices including irrigation of crops that also contribute to increased number of microorganisms in water.

Bacteria are found in the shores and depths of the seas including the hypersaline dead sea with salinity above 35%o (parts per thousand). These are the extremely halophilic bacteria e.g. *Halobacterium halobium* (Haruhico *et al.*(1995). Research on halophiles has contributed to understanding the activities of osmophilic bacteria in the preservation of food with salt or sugar. Microbes get adapted to environmental changes or challenges by the modification of their metabolic pathways and evolution of genetic apparatus required to meet the demands of the new situation. This knowledge is exploited by man in the production of microbial metabolites like antibiotics, and in genetic engineering of microbes for specific characters.

The isolation of microbes from aquatic thermal vents opened a new vista for producing industrial enzymes that are stable at high temperatures (100?C).

Therefore, denaturation of enzymes is avoided and the cost of cooling to protect mesophilic enzymes is saved. Eveleigh and others at Cook College of Rutgers University, USA elucidated the enzyme of an extremophilic bacterium. I had the privilege of working with Douglas Eveleigh during my last sabbatical leave.

Water also occurs underground in water tables or aquifers having percolated through the soil layers. Hence, spring water arising from the aquifer is ordinarily clear and palatable. However, seepage of effluents in refuse dumps and other human activities involving the subsurface of the earth aggravate the contamination of ground water thus interfering with the quality of spring water. Microorganisms in ground water constitute a health hazard to consumers. In a setting with little or irregular supply of pipe-borne water and usually, well water quality is not tested, consumers are vulnerable to waterborne infections.

It is the scarcity of potable water that has enhanced the sale of the so-called 'pure 'water sachets. For safety of packaged drinking water in Nigeria, the activities of the National Agency for Food Drug Administration Control (NAFDAC) deserve commendation. As an environmentalist, I would add that the used sachets constitute a nuisance especially as the waste is recalcitrant, and being non-biodegradable, it is persistent in the environment. Furthermore, care needs to be taken about the location of disposal sites for solid wastes.

The geology of the underlying soil should be considered and the disposal site may be lined before dumping, and the refuse covered with an impervious material. These efforts would control wetting of refuse by rainfall and mitigate the seepage of effluents into the aquifer with its eventual contamination and negative impact on consumers.

Water Pollution

The aquatic environment has a natural ability for self-purification. Physicochemical and biologic activities including microbiologic action bring about the breakdown of small quantities of organic materials. In effect, contaminated body of water such as a stream can be cleansed through selfpurification processes (Akpata, 1998).

It is when the type of waste is nonbiodegradable e.g. plastics or the amount of organic input exceeds the carrying capacity of the environment that there emanate undesirable conditions that elicit negative impacts on the ecosystem. This is the stage that there is environmental pollution.

Faecal Pollution of Lagos Lagoon

A trip using the famous UNILAG 1, motorboat of the erstwhile Department of Biological Sciences, was undertaken to survey the Lagos lagoon for areas of point source pollution. Nobody could ignore Iddo site of faeces disposal for the stench and unaesthetic value. There was a need to enquire from the Lagos City Council just what was going on? Iddo is located on the Lagos Mainland and is accessible to Mushin, Yaba, Ebute- Metta, Surulere, and so on. The faeces point was near the Carter Bridge on the mainland end. The rationale for disposal of raw faeces into the Lagos lagoon at Iddo was that the waste would be washed into the sea and diluted to extinction. In that case, the waste would not be seen and there would be no visible or apparent impact on the aquatic ecosystem.

To clarify the situation, a research study was designed to investigate the distribution of faeces in the lagoon. Samples of fresh faeces obtained from the disposal vehicle or lagoon water were cultured for pathogens and commensals. The Lagos City Council recorded the volumes of faeces deposited into the lagoon. Seven sampling stations were identified on the lagoon, some were located to the north while others were to the south of the faeces disposal point. An apparently unpolluted station located in the middle of the lagoon served as control. Some physicochemical parameters e.g. temperature, pH, salinity, suspended organic matter and turbidity were measured at each sampling station. Faecal pollution was assayed using faecal indicator bacteria namely, coliforms, *Escherichia coli, Streptococcus faecalis or Clostridium perfringes* for their value in determining the period of pollution. Vertical and horizontal distribution of faecal bacteria in the lagoon was studied at each sampling station so as to determine the abundance and distribution of faeces in the aquatic ecosystem.

The results showed presence of pathogens e.g. *Salmonella* and *Shigella* species in the faeces samples from some areas. However, pathogens were not isolated from stations on the lagoon. It was observed that *Escherichia coli* count decreased rapidly southwards near the sea and less rapidly northwards nearer Ogun River (Akpata & Ekundayo, 1978). It is probable that stations where faecal bacteria were isolated had been contaminated with faeces and possibly pathogens. There was both horizontal and vertical distribution of faeces in the lagoon. Higher counts of bacteria were obtained in bottom samples an indication that sedimentation of faeces occurred at each station. This could be due to the effect of gravity and increased organic nutrients. Twenty-six million litres of faeces were deposited into the lagoon at Iddo in 1973. Our results showed that the faeces were not always washed southwards into the sea from the Lagos lagoon as suggested by the Lagos City Council.

WATER MICROBIOLOGY

Water microbiology is concerned with the microorganisms that live in water, or can be transported from one habitat to another by water.

Water can support the growth of many types of microorganisms. This can be advantageous. For example, the chemical activities of certain strains of yeasts provide us with beer and bread. As well, the growth of some bacteria in contaminated water can help digest the poisons from the water.

However, the presence of other disease causing microbes in water is unhealthy and even life threatening. For example, bacteria that live in the intestinal tracts of humans and other warm blooded animals, such as *Escherichia coli, Salmonella, Shigella,* and *Vibrio,* can contaminate water if feces enters the water. Contamination of drinking water with a type of *Escherichia coli* known as O157:H7 can be fatal. The contamination of the municipal water supply of Walkerton, Ontario, Canada in the summer of 2000 by strain O157:H7 sickened 2,000 people and killed seven people.

The intestinal tract of warm-blooded animals also contains viruses that can contaminate water and cause disease. Examples include rotavirus, enteroviruses, and coxsackievirus.

Another group of microbes of concern in water microbiology are protozoa. The two protozoa of the most concern are *Giardia* and *Cryptosporidium*. They live normally in the intestinal tract of animals such as beaver and deer. *Giardia* and *Cryptosporidium* form dormant and hardy forms called cysts during their life cycles. The cyst forms are resistant to chlorine, which is the most popular form of drinking water disinfection, and can pass through the filters used in many water treatment plants. If ingested in drinking water they can cause debilitating and prolonged diarrhea in humans, and can be life threatening to those people with impaired immune systems. *Cryptosporidium* contamination of the drinking water of Milwaukee, Wisconsin with in 1993 sickened more than 400,000 people and killed 47 people. Many microorganisms are found naturally in fresh and saltwater. These include bacteria, cyanobacteria, protozoa, algae, and tiny animals such as rotifers. These can be important in the food chain that forms the basis of life in the water. For example, the microbes called cyanobacteria can convert the energy of the sun into the energy it needs to live. The plentiful numbers of these organisms in turn are used as food for other life. The algae that thrive in water is also an important food source for other forms of life. A variety of microorganisms live in fresh water. The region of a water body near the shoreline (the littoral zone) is well lighted, shallow, and warmer than other regions of the water. Photosynthetic algae and bacteria that use light as energy thrive in this zone. Further away from the shore is the limnitic zone. Photosynthetic microbes also live here. As the water deepens, temperatures become colder and the oxygen concentration and light in the water decrease. Now, microbes that require oxygen do not thrive. Instead, purple and green sulphur bacteria, which can grow without oxygen, dominate. Finally, at the bottom of fresh waters (the benthic zone), few microbes survive.

Bacteria that can survive in the absence of oxygen and sunlight, such as methane producing bacteria, thrive.

Saltwater presents a different environment to microorganisms. The higher salt concentration, higher pH, and lower nutrients, relative to freshwater, are lethal to many microorganisms. But, salt loving (halophilic) bacteria abound near the surface, and some bacteria that also live in freshwater are plentiful (i.e., *Pseudomonas* and *Vibrio*). Also, in 2001, researchers demonstrated that the ancient form of microbial life known as archaebacteria is one of the dominant forms of life in the ocean. The role of archaebacteria in the ocean food chain is not yet known, but must be of vital importance. Another microorganism found in saltwater are a type of algae known as dinoflagellelates. The rapid growth and multiplication of dinoflagellates can turn the water red. This "red tide" depletes the water of nutrients and oxygen, which can cause many fish to die. As well, humans can become ill by eating contaminated fish. Water can also be an ideal means of transporting microorganisms from one place to another. For example, the water that is carried in the hulls of ships to stabilize the vessels during their ocean voyages is now known to be a means of transporting microorganisms around the globe. One of these organisms, a bacterium called *Vibrio cholerae*, causes life threatening diarrhea in humans. Drinking water is usually treated to minimize the risk of microbial contamination. The importance of drinking water treatment has been known for centuries. For example, in pre-Christian times the storage of drinking water in jugs made of metal was practiced. Now, the anti-bacterial effect of some metals is known. Similarly, the boiling of drinking water, as a means of protection of water has long been known. Chemicals such as chlorine or chlorine derivatives has been a popular means of killing bacteria such as *Escherichia coli* in water since the early decades of the twentieth century. Other bacteria-killing treatments that are increasingly becoming popular include the use of a gas called ozone and the disabling of the microbe's genetic material by the use of ultraviolet light. Microbes can also be physically excluded form the water by passing the water through a filter. Modern filters have holes in them that are so tiny that even particles as miniscule as viruses can be trapped.

An important aspect of water microbiology, particularly for drinking water, is the testing of the water to ensure that it is safe to drink. Water quality testing can de done in several ways. One popular test measures the turbidity of the water. Turbidity gives an indication of the amount of suspended material in the water. Typically, if material such as soil is present in the water then microorganisms will also be present. The presence of particles even as small as bacteria and viruses can decrease the clarity of the water. Turbidity is a quick way of indicating if water quality is deteriorating, and so if action should be taken to correct the water problem. In many countries, water microbiology is also the subject of legislation. Regulations specify how often water sources are sampled, how the sampling is done, how the analysis will be performed, what microbes are detected, and the acceptable limits for the target microorganisms in the water sample. Testing for microbes that cause disease (i.e., *Salmonella typhymurium* and *Vibrio cholerae*) can be expensive and, if the bacteria are present in low numbers, they may escape detection. Instead, other more numerous bacteria provide an indication of fecal pollution of the water. *Escherichia coli* has been used as an indicator of fecal pollution for decades. The bacterium is present in the intestinal tract in huge numbers, and is more numerous than the disease-causing bacteria and viruses. The chances of detecting *Escherichia coli* is better than detecting the actual disease causing microorganisms. *Escherichia coli* also had the advantage of not being capable of growing and reproducing in the water (except in the warm and food-laden waters of tropical countries). Thus, the presence of the bacterium in water is indicative of recent fecal pollution. Finally, *Escherichia coli* can be detected easily and inexpensively.

There are two major types of water.

1. Ground water - It originates from deep wells and subterranean springs. This is virtually free of bacteria due to filtering action of soil, deep sand and rock. However, it may become contaminated when it flows along the channels.
2. Surface water - It is found in streams, lakes, and shallow wells. The air through which the rain passes may contaminate the water. Other sources are the various types of establishments and agricultural farms etc. by the sides the water flows.

Possible sources of microbial contamination of a body of water are soil and agricultural run off, farm animals, rain water, industrial waste, discharges from sewage treatment plants and storm water run off from urban areas. In water microbiology the water is contaminated when it contains a chemical or biological poison or an infectious agent. These conditions also apply to water which is polluted except that the agent or poison is often obvious and the water carries an unpleasant taste or appearance. Potability refers to the drinkability of water. When potable, it is fit for drinking. When unpotable it is unfit due to some contaminant or pollutant.

ATMOSPHERIC WATERS

Rain, snow and hail which fan on land tend to carry down particles of dust, soot, and other materials suspended in the air. These often bear bacteria and other microorganisms -on their surface. The number of organisms depends upon local conditions. After heavy rain or snow the atmosphere is washed free of organisms. Surface Waters - As soon as ram or snow reaches the earth and flows over the soil, some of the soil organisms are gathered up by the water. Bodies, of water such as streams, rivers, and oceans represent surface water.

Microbial populations depend upon their numbers in the soil and, also, upon the kinds and quantities of food material dissolved out of the soil by water. Climatic, geographical and biological conditions bring about great variations in microbial populations of surface waters.

Rivers and streams show their highest count during the rainy period Dust blowing into rivers and streams also contributes many microorganisms. Animals also make considerable contribution to the microbial flora of the surface waters. They bathe and often drop their excreta in the water.

Stored Waters - Inland waters held in ponds, lakes or reservoirs represent stored waters. Storage generally reduces the numbers of organisms in water. A certain degree (If purity and stability is established.

Several factors affect the microbial flora of stored waters. These are as follows;

Sedimentation

Microorganisms have a specific gravity slightly greater than that of' water, and therefore slowly settle down. However, the most important factor is their attachment to suspended particles. Microorganisms are removed from the upper layers of the water as the suspended particles settle down.

Activities of other Organisms

Predatory Protozoa engulf living or dead bacteria for food, provided the water contains sufficient dissolved oxygen.

Light ray: Direct sunlight is toxic to both vegetative cells and spores of microorganisms.

The toxicity of ultraviolet rays is inversely proportional to the turbidity of water. In tropical countries direct sun light is a very effective sterilizing agent.

Temperature

Temperature has variable effects. It may kill some organisms and may stimulate the growth of others. During colder months the multiplication rate of microorganisms is considerably reduced.

Food Supply

If there is considerable vegatation or suspended food particles in the body of water, it is likely to increase the number of organisms. On the other hand, certain toxic substances may bring about marked reduction in the number of, organisms.

Ground Waters

As water seeps through the earth, microorganisms as well as suspended particles are removed by filtration in varying degrees. This depends on the permeability characteristics of the soil and the depth to which the water penetrates.

Ground water brought to the surface by springs or deep wells contains very few organisms. To construct a well the nature of the soil and underlying porous strata, the nature of the water table, and the nature, distance, and direction of the local sources of pollution must be taken into consideration.

Faecal Contamination of Water

Water gets contaminated with pathogenic microorganisms through intestinal discharges of man and animals. Further more, in the intestinal tract of man and animals, there exists a characteristic group of organism's designated as coliforms.

The coliform group of bacteria includes aerobic and facultatively anaerobic, gram negative, nonsporeforming bacilli which ferment lactose with acid and produce gas within 48 hours at, 35°C. The most common species of this group are various strains of Escherichia coli and Aerobacter aerogenes.

E. coli is commonly found in the intestinal tract of man and animals, while A. aerogenes is normally found on plants and grains, and may sometimes occur in the intestinal tract of man and animals. Contamination of water with either type makes the water unsatisfactory for drinking purposes.

Marine Microbiology

The study of microorganisms living in the sea is known as marine microbiology. Nearly three fourths of the earth's surface is covered by the sea.The sea therefore is the largest natural environment inhabited by the microorganisms. The marine environment, as does land, contains bacteria, Protozoa, algae, yeasts and moulds, as well as viruses forms organisms that live in the sea.

Microorganisms which are found in the surface layers of the sea and other aquatic environment are collectively known as plankton.

Plankton may be further sub divided into phytoplank-ton (photosynthetic microorganisms, algae) and zooplankton (Protozoa, and other microscopic animals). The former are more important plankton organisms, as they arc the primary producers of organic matter.

Bacteria and fungi are also present in large numbers in the plankton. Microorganisms present in the bottom regions (sediments) arc designated the benthos or benthic community. A variety of microorganisms arc found in this region. Bacteria, however, predominate.

The microbial flora characteristic of tile sea has one factor in its environmental pattern which differs from freshwater lakes, streams, and rivers, for the sea is salty. There is a high concentration of salt and mineral ions in the sea. The principal salts are the chlorides, sulphates, and carbonates of sodium, potassium, calcium and magnesium.

The concentration of dissolved salts (salinity) averages about 35 gin/kg or 3.5% by weight. Out of this 75% is NaCl. Marine microorganisms, there-fore; may be divided into two general types:

(1) those indigenous to the sea and not growing on media without sea water or high salt concentra-tion and
(2) transient organisms whose natural habitat is terrestrial and which are able to grow in media without sea water, but can tolerate high salt concentration.

There is also another group of microorganisms in the sea, in the depths of the sea microorganisms live at tremendous hydrostatic pressures, upto 1000 atmospheres. These organisms when brought to the surface require not only sea water but high pressure in order to grow. Such pressures are toxic to organisms normally living at one atmosphere at the surface.

Bacteriological Examination of Domestic Water

Natural water supplies such as rivers, lakes, and streams contain sufficient nutrients to support growth of various organisms. Microorganisms enter the water supply in several different ways. In congested centers water supplies get polluted by domestic and industrial wastes, As a potential carrier of pathogenic microorganisms, water can endanger health and life.

From the standpoint of transmitting human diseases, polluting waters with soil, rubbish, industrial wastes, and even animal manure is comparatively harmless. These sources rarely contain pathogens capable of producing human diseases when swallowed with drinking water. Sewage containing human excreta, however, is the most dangerous material that pollutes water.

People with communicable diseases of many kinds eliminate the causative organisms in their excreta. The most important microbial diseases transmitted through water are typhoid fever, paratyp-hoid fever, amoebic dysentery, bacillary dysentery, cholera, tularemia, poliomyelitis, and infectious hepatitis.

To determine the potability of water quantitative bacteriological examination may be undertaken. However, there is no single test or, even combination of tests, that is wholly satisfactory, because it will give only a fraction of the total count. Theoretically it would be better to examine water for the presence of the specific pathogenic microorganisms. This is also impracticable because of the following reasons:

1. The methods are expensive, tedious, and slow, and by that time the water has already been consumed.
2. The number of pathogenic organisms may be quite small compared to non pathogenic organisms and would be overlooked.
3. Non-pathogenic organisms may interfere with the examination of pathogens.

The direct examination for pathogens, therefore, is not used in routine water analysis. Methods commonly used for the bacteriological examination of water, are based on:

1. The examination of presence or absence of the more common organisms of intestinal or sewage origin.
2. The approximate determination of total numbers of bacteria present in the water sample.

Microbial Components of Water

The type of microorganisms differ in unpolluted, polluted and marine waters.. In unpolluted water of mountain lake or stream there are usually low organic nutrients.

The number of bacteria is very much limited, a few thousand per ml. Actinomycetes are typical. Autotrophic bacteria are common alongwith free living protozoans as Euglena, Paramecium and various amoebae.

Marine Water

The marine water is high in salt and only halophilic organisms survive. Since temperature is very low, most of them are psychrophilic. Diatoms and protozoa as dinoflagellates are important components in food chain. In the off shore oceanic zone, photosynthetic organisms are present. These are diatoms and dinoflagellates.

Most marine microbes are found along the shoreline or littoral zone where nutrients are in abundance. Some unusual types are also present on the ocean floor in the benthic zone and even at the bottom of several miles-deep trenches, the abyssal zone.

Polluted Water

In polluted waters, there are large amounts of organic matter from sewage, feces and industrial complex. The microbes are usually heterotrophic. The digestion of organic matter by these organisms is incomplete, due to which there accumulate acids, bases, alcohols and various gases.

The major type of bacteria are coliform bacteria the Gram-negative nonsporeforming bacilli usually found in the intestine. This group includes E. coli and species of Enterabacter. They ferment lactose to acid and gas. Noncoliform bacteria Streptococcus, Proteus and Pseudomonas are also present.

Under some conditions, the polluting organisms multiply rapidly and consume most of the available oxygen. For instance, nutrients enter the river from sources like sewage treatment plants or urban suburban runoff.

Thus river suddenly develops a high nutrient content. Under these conditions algae may bloom rapidly. This leads to depletion of oxygen in water. There is very little oxygen available to the protozoa, small animals, fish and plants. Due to this non-availability of oxygen, a layer of dead organisms, mud and silt accumulate at the bottom, and anaerobeic species of Clostridium, Desulfovibrio etc. will flourish.

They produce gases. One gas, H2S combines with lead or iron to give a precipitate which makes the mud black and the water poisonous. Due to complete depletion of oxygen, the suspended bacteria die in their own waste products. There is hardly any life in water at this stage. The gas bubbles from the anaerobes in the mud break the surface. Such processes lead to death of a river.

Water Pollution

Polluted waters can be classified in different ways. In microbiological context, there are generally recognised physical, chemical and biological pollution of water.

Physical Water Pollution

Here the water becomes cloudy due to presence of some particulate matter such as sand or soil. Some plant material as algal-bloom may also be involved, that turns the consistency of water like a pea soup. Presence of some phosphates or nitrates may encourage the growth of blooms leading finally to a condition known as eutrophication. This upsets the ecological balance.

Chemical Water Pollution

It occurs due to introduction of inorganic and organic waste to water. For example, water passing out of a mine contains large amount of copper, iron or other ore. Another source of pollution is metal pipelines. Corrosion of metals of the pipes may yield rusty or black precipitates.

Other chemical pollutants include laundry detergents, radioactive wastes and nitrates from fertilisers etc. An important group of chemicals are polychlorinated biphenyls or PCBs, a group of industrial chemicals used in the production of electric insulators and paper products. They are carcinogenic and necessary to be eliminated from water. Main sources of chemical pollutants are mining wastes, air pollutants (washed down by rain), power plants, industries, sewage and storm water runoff, wash from farm lands (pesticides, fertilisers), hospitals, road salts and exhausts from automobiles.

Biological Water Pollution

It develops from the microorganisms that enter waters from such sources as human waste, food operations, meat packing plants and medical facilities etc. Under normal conditions, the water body is able to handle biological material, because heterotrophs digest the organic matter to simple phosphates, nitrates and sulphates which fertilise the plants.

Carbohydrates are converted to CO2 and water, which are also useful to plants. The plants then release °2 through photosynthesis. Due to rapid movement of water, aeration is more or less constant and the wastes eliminated.

However, there arises the problem when water becomes stagnant or is overloaded with waste. Water is now unable to handle the biological material, and it soon becomes polluted. We have already seen how a lake may turn into a swamp due to overabundance of organic matter.

A critical factor operating under such a situation is the biological oxygen demand (BOD). It refers to the oxygen requirement by the metabolising organisms. As the number of organisms increases, the demand for O2 increases proportionally. Depletion of O2 in water may lead to death of fish and other aquatic animals, a flat taste of water, and eventual death of aerobes.

The BOD is therefore an important indication of the levels of biological pollution in the water. In laboratory, BOD is determined by measuring the amount of dissolved oxygen in test samples of water distributed to BOD bottles. Initial determination of dissolved oxygen in sample is made.

The bottles are then stoppered and incubated at 20°C for five days, after which the amount of dissolved oxygen is again determined. The difference in the oxygen content represents the amount of oxygen used up in the test sample, which is the BOD. Results are expressed as parts per million (ppm). A BOD of several hundred is taken usually high.

Diseases Transmitted by Water

Water is infact a vehicle for the transfer of a wide range of diseases of microbial origin. The important bacterial diseases include typhoid fever, cholera and bacterial dysentry that are generally transmitted when human feces from carriers or patients contaminates the water. Viral diseases transmitted by water include virus A hepatitis and polio. Many protozoa form cysts which survive for long periods in water. The causal agents of amoebiasis, giardiasis etc. are important concerns in water pollution.

Water Treatment

We have seen that water and sewage harbour a variety of organisms responsible for diseases in man. It is thus necessary to remove or rather interrupt the disease cycle in nature through treatment of water and sewage. Water purification and sanitary sewage disposal are two major methods for interrupting disease cycle. In water purification organisms are prevented from reaching the body whereas in sewage disposal, they are removed from body waste products. The various steps in the purification of drinking water in municipal water supplies are, It may be seen that there are three basic steps in the purification of drinking water: sedimentation, filtration and chlorination.

Sedimentation Process

Water usually undergoes some degree of purification during storage in ponds or reservoirs. Suspended particles settle and carry down most of the microorganisms. The rate of purification by sedimentation depends upon the kind and amount of suspended matter as the well as physical, chemical and biological conditions of the stored water. The rate of sedimentation is enhanced by adding alum, iron salts, colloidal silicate, etc., which produce flocculent precipitates. Microorganisms and suspended particles are entrapped and settle rapidly. Sometimes activated carbon is also added. This adsorbs the compound responsible for objectionable colour and taste of water. Microorganisms remain viable for a considerable time, even though visible evidence of pollution has disappeared. Sedimentation, therefore, reduces the microbial population but does not sterilize polluted water. To produce potable water further treatment is necessary. Thus sedimentation is often used as a first stage purification.

Filtration Process

This process removes the remaining microorganisms from the water. Although, different types of filter material are available, mostly a layer of sand and gravel is utilised to trap the microorganisms. A slow sand filter uses several feet deep layer of fine sand particles. Within the sand there forms a layer of microorganisms which acts as an additional filter. This layer is called a Schmutzdecke or dirty layer. To cleanse the filter, the top layer is removed and replaced with fresh sand. A rapid sand filter contains coarser particles of gravel. The schmutzdecke does not develop; the filtration rate is higher (over 200 million gallons per acre per day). This type of filter is generally used in municipal supplies.

Slow Sand Filter

Slow sand filtration plants require considerable area because the rate of filtration is slow. A concrete floor with drainage tiles to collect the filtered water is constructed. The tile is covered with coarse gravel, fine gravel, coarse sand and finally 2 to 1 feet of sand at the top. Water seeps through the filter slowly., is collected by tile drain pipes at the bottom, and is pumped into a reservoir. At best five million gallons of water per acre, per day, can be filtered. Slow sand filters are clogged by turbid water. Water to be filtered is, therefore, clarified by sedimentation with or without coagulation. The purification of water is accomplished not by the screening action of the sand for the spaces arc much to large, but by a different principle. A colloidal, flocculent material composed of bacteria, algae, and Protozoa accumulates in the surface layers of fine sand. This Slimy, gelatinous film closes up the pores between the sand grains and makes the filter bed more and more effective. Since bacteria have a negative electrical charge and colloidal material on the sand grains has a. positive charge, bacteria are thus adsorbed on the particles.

Bacteria arc also injested by Protozoa that Inhabit the upper layer of the film. Metabolic activity of microorganisms also greatly reduces the chemical content of the water.

When, the gelatinous film finally become too thick, the efficiency of the filter gradually decreases. The filter is taken out of service and the surface layer is cleaned.

Rapid Sand Filter

Rapid sand filters are constructed in a manner similar to that of slow sand filters. They also consist of layers of sand, gravel, and rock. Water is pretreated before filtration by a coagulant such as alum or ferrous sulphate.

The water passes through a settling tank in which most of the precipitate settles out, and the remainder is pumped on to the filter. Rapid sand filters soon become clogged and are cleaned by forcing cleaned water backward (back washed) through the bed of gravel and sand, and bubbling air through them.

The back water rises through the filter and carries the accumulated material to the sewer. The wash water is thus wasted, Care is taken in this backwashing procedure to see that the fine sand on the surface is not lost.

Rapid sand filters are usually operated in batteries, so that some may be in operation, while others are being cleaned.

They are nearly as effective as slow sand filters but operate 50 times faster than slow sand filters. Rapid sand filters are capable of delivering 150 to 200 million gallons of water per acre, per day. They require a much smaller area of land for more water filtrations and cost much less to install and maintain.

Many other filtration devices such as pressure filters, diatomite fitters, membrane filter, reverse osmosis etc, are employed to remove various impurities in water. Recovery of potable water from the sea and from domestic and. industrial sewage is also undertaken by the use of filtration techniques.

Chlorination Process

In this final step, chlorine gas is added to the water. This gas is an active oxidant which reacts with organic matter in the water. Gas is added until any residue is present. A residue of 0.2 ppm-1.0 ppm of water is often the standard used. Under these conditions, most remaining microbes die within 30 minutes.

Bacteriological Analysis of Water

There are methods to detect bacterial contamination of water. The chief objective is to identify coliform organisms as Escherichia coli. Their presence indicates that water contains fecal pollution and is unsafe for consumption.

Membrane Filter Technique

It may be used in the field also. A special, collecting bottle is held against the water current and a 100-ml sample is taken. The water is filtered through a membrane filter and the filter pad is then transferred to a plate of bacteriological medium. Bacteria trapped in the filter will form colonies which may be counted.

Standard Plate Count SPC

Water sample is diluted in sterile buffer. Measured amounts are pipetted into Petri dishes. Agar medium is added and plates incubated. Colony counts are made and multiplied by the reciprocal dilution factor to have total bacteria per ml water. This method is similar to that used for milk.

Most Propable Number

In this procedure, water in 10 ml, I mi and 0.1 mt amounts is inoculated into lactose broth tubes. The tubes are incubated and coliform organisms may be identified by their production of gas from the lactose.

By referring to a MPN table, a statistical range of the number of coliform may be determined by observing how many broth tubes showed gas.

This method does not detect total number of bacteria in the water nor it locate noncoliforms like Salmonella. However it indicates the presence and quantity of coliforms.

Specific Tests For Asseing Water Coliform Counts

The most frequently used indicator organism is the normally nonpathogenic Coliform bacterium Escherichia coli. Positive tests for E. coli do not prove the presence of enteropathogens, such as Salmonella and Shigella, but do establish the possibility. E. coli is more numerous and easier to be grown.

Presumptive Test

A series of lactose broth or lauryl sulphate tryptose broth fermentation tubes are inoculated with measured amounts of water and incubated at 35°C for 24 to 48 hours. The formation of gas in the inverted vial in the fermentation tube within 48 hours indicates a positive presumptive test. Absence of gas formation at the end of 48 hours constitutes a negative presumptive test. This means that the water sample does not contain coliforms and is considered safe.

Confirmed Test

Sometimes a false positive presumptive test is obtained. This may be due to the presence of yeasts, certain Clostridium species and some other organisms, or by bacterial associations or synergism. In order to be certain that gas-production is due to coliforms a confirmed test must be performed. Two procedures are normally employed. In one method a drop of culture from a positive lactose broth tube is transferred to brilliant green lactose bile fermentation broth, and is incubated for 24 to 48 hours at 35°C. The appearance of gas within 48 hrs constitues a positive confirmed test. The dye inhibits gram positive organisms and eliminates a false presumptive test and the syngergistic reaction of gram positive and gram negative organisms growing together. In the second method a drop of culture from the positive lactose broth is streaked on a petriplate containing, Endo's, or eosin-methylene blue agar. The appearance of nucleated colonies, with or without a metallic sheen, within 24 hours indicates a positive confirmed test.

Completed Test

Isolated colonies from petriplates are transferred into lactose fermentation broth and streaked on to an agar slant. The presence of gas in the fermentation broth and the presence of gram negative non spore-forming bacilli on the slant give evidence that coliform bacteria were present in the original water sample.

IMViC Reaction

E. coli and A. aerogenes are normally referred to as faecal and non-faecal contaminants of water, respectively, and are the most important organisms of the coliform group. Since they closely resemble each other in their morphological and cultural characteristics, biochemical tests, are, performed to differentiate them. These tests are collectively designated as the IMViC reactions. The name was coined by Parr from the first letters of the four tests, namely Indole, Methyl red, Voges Proskauer, and Citrate. There are 16 possible combinations of positive and negative tests of these four characteristics. Most, of these combinations have been found, but the reactions of E. coli and A. aerogenes are commonly found. The remaining 14 types are usually designated as "intermediates".

ENVIRONMENTAL BIOSPHERE

The biosphere is the region of the earth that encompasses all living organisms: plants, animals and bacteria. It is a feature that distinguishes the earth from the other planets in the solar system. "Bio" means life, and the term biosphere was first coined by a Russian scientist (Vladimir Vernadsky) in the 1920s. Another term sometimes used is ecosphere ("eco" meaning home). The biosphere includes the outer region of the earth (the lithosphere) and the lower region of the atmosphere (the troposphere).

It also includes the hydrosphere, the region of lakes, oceans, streams, ice and clouds comprising the earth's water resources. Traditionally, the biosphere is considered to extend from the bottom of the oceans to the highest mountaintops, a layer with an average thickness of about 20 kilometers. Scientists now know that some forms of microbes live at great depths, sometimes several thousand meters into the earth's crust.

Nonetheless, the biosphere is a very tiny region on the scale of the whole earth, analogous to the thickness of the skin on an apple. The bulk of living organisms actually live within a smaller fraction of the biosphere, from about 500 meters below the ocean's surface to about 6 kilometers above sea level.

Dynamic interactions occur between the biotic region (biosphere) and the abiotic regions (atmosphere, lithosphere and hydrosphere) of the earth. Energy, water, gases and nutrients are exchanged between the regions on various spatial and time scales. Such exchanges depend upon, and can be altered by, the environments of the regions. For example, the chemical processes of early life on earth (e.g. photosynthesis, respiration, carbonate formation) transformed the reducing ancient atmosphere into the oxidizing (free oxygen) environment of today. The interactive processes between the biosphere and the abiotic regions work to maintain a kind of planetary equilibrium. These processes, as well as those that might disrupt this equilibrium, involve a range of scientific and socioeconomic issues. The study of the relationships of living organisms with one another and with their environment is the science known as ecology. The word ecology comes from the Greek words oikos and logos, and literally means "study of the home." The ecology of the earth can be studied at various levels: an individual (organism), a population, a community, an ecosystem, a biome or the entire biosphere. The variety of living organisms that inhabit an environment is a measure of its biodiversity.

ORGANISMS

Life evolved after oceans formed, as the ocean environment provided the necessary nutrients and support medium for the initial simple organisms. It also protected them from the harsh atmospheric UV radiation. As organisms became more complex they eventually became capable of living on land. However, this could not occur until the atmosphere became oxidizing and a protective ozone layer formed which blocked the harmful UV radiation. Over roughly the last four billion years, organisms have diversified and adapted to all kinds of environments, from the icy regions near the poles to the warm tropics near the equator, and from deep in the rocky crust of the earth to the upper reaches of the troposphere. Despite their diversity, all living organisms share certain characteristics: they all replicate and all use DNA to accomplish the replication process. Based on the structure of their cells, organisms can be classified into two types: eukaryotes and prokaryotes. The main difference between them is that a eukaryote has a nucleus, which contains its DNA, while a prokaryote does not have a nucleus, but instead its DNA is free-floating in the cell. Bacteria are prokaryotes, and humans are eukaryotes. Organisms can also be classified according to how they acquire energy. Autotrophs are "self feeders" that use light or chemical energy to make food. Plants are autotrophs. Heterotrophs (i.e. other feeders) obtain energy by eating other organisms, or their remains. Bacteria and animals are heterotrophs. Groups of organisms that are physically and genetically related can be classified into species. There are millions of species on the earth, most of them unstudied and many of them unknown. Insects and microorganisms comprise the majority of species, while humans and other mammals comprise only a tiny fraction. In an ecological study, a single member of a species or organism is known as an individual.

POPULATIONS AND COMMUNITIES

A number of individuals of the same species in a given area constitute a population. The number typically ranges anywhere from a few individuals to several thousand individuals. Bacterial populations can number in the millions. Populations live in a place or environment called a habitat. All of the populations of species in a given region together make up a community. In an area of tropical grassland, a community might be made up of grasses, shrubs, insects, rodents and various species of hoofed mammals. The populations and communities found in a particular environment are determined by abiotic and biotic limiting factors. These are the factors that most affect the success of populations. Abiotic limiting factors involve the physical and chemical characteristics of the environment. Some of these factors include: amounts of sunlight, annual rainfall, available nutrients, oxygen levels and temperature. For example, the amount of annual rainfall may determine whether a region is a grassland or forest, which in turn, affects the types of animals living there. Each population in a community has a range of tolerance for an abiotic limiting factor. There are also certain maximum and minimum requirements known as tolerance limits, above and below which no member of a population is able to survive. The range of an abiotic factor that results in the largest population of a species is known as the optimum range for that factor. Some populations may have a narrow range of tolerance for one factor. For example, a freshwater fish species may have a narrow tolerance range for dissolved oxygen in the water. If the lake in which that fish species lives undergoes eutrophication, the species will die. This fish species can therefore act as an indicator species, because its presence or absence is a strict indicator of the condition of the lake with regard to dissolved oxygen content.

Biotic limiting factors involve interactions between different populations, such as competition for food and habitat. For example, an increase in the population of a meat-eating predator might result in a decrease in the population of its plant-eating prey, which in turn might result in an increase in the plant population the prey feeds on. Sometimes, the presence of a certain species may significantly affect the community make up. Such a species is known as a keystone species. For example, a beaver builds a dam on a stream and causes the meadow behind it to flood. A starfish keeps mussels from dominating a rocky beach, thereby allowing many other species to exist there.

Ecosystems

An ecosystem is a community of living organisms interacting with each other and their environment. Ecosystems occur in all sizes. A tidal pool, a pond, a river, an alpine meadow and an oak forest are all examples of ecosystems. Organisms living in a particular ecosystem are adapted to the prevailing abiotic and biotic conditions. Abiotic conditions involve both physical and chemical factors (e.g., sunlight, water, temperature, soil, prevailing wind, latitude and elevation). In order to understand the flow of energy and matter within an ecosystem, it is necessary to study the feeding relationships of the living organisms within it.

Living organisms in an ecosystem are usually grouped according to how they obtain food. Autotrophs that make their own food are known as producers, while heterotrophs that eat other organisms, living or dead, are known as consumers. The producers include land and aquatic plants, algae and microscopic phytoplankton in the ocean. They all make their own food by using chemicals and energy sources from their environment.

For example, plants use photosynthesis to manufacture sugar (glucose) from carbon dioxide and water. Using this sugar and other nutrients (e.g., nitrogen, phosphorus) assimilated by their roots, plants produce a variety of organic materials. These materials include: starches, lipids, proteins and nucleic acids. Energy from sunlight is thus fixed as food used by themselves and by consumers.

The consumers are classed into different groups depending on the source of their food. Herbivores (e.g. deer, squirrels) feed on plants and are known as primary consumers. Carnivores (e.g. lions, hawks, killer whales) feed on other consumers and can be classified as secondary consumers. They feed on primary consumers. Tertiary consumers feed on other carnivores. Some organisms known as omnivores (e.g., bears, rats and humans) feed on both plants and animals. Organisms that feed on dead organisms are called scavengers (e.g., vultures, ants and flies). Detritivores (detritus feeders, e.g. earthworms, termites, crabs) feed on organic wastes or fragments of dead organisms.

Decomposers (e.g. bacteria, fungi) also feed on organic waste and dead organisms, but they digest the materials outside their bodies. The decomposers play a crucial role in recycling nutrients, as they reduce complex organic matter into inorganic nutrients that can be used by producers. If an organic substance can be broken down by decomposers, it is called biodegradable.

In every ecosystem, each consumer level depends upon lower-level organisms (e.g. a primary consumer depends upon a producer, a secondary consumer depends upon a primary consumer and a tertiary consumer depends upon a secondary consumer).

All of these levels, from producer to tertiary consumer, form what is known as a food chain. A community has many food chains that are interwoven into a complex food web. The amount of organic material in a food web is referred to as its biomass.

When one organism eats another, chemical energy stored in biomass is transferred from one level of the food chain to the next. Most of the consumed biomass is not converted into biomass of the consumer. Only a small portion of the useable energy is actually transferred to the next level, typically 10 percent. Each higher level of the food chain represents a cumulative loss of useable energy. The result is a pyramid of energy flow, with producers forming the base level. Assuming 10 percent efficiency at each level, the tertiary consumer level would use only 0.1 percent of the energy available at the initial producer level. Because there is less energy available high on the energy pyramid, there are fewer top-level consumers. A disruption of the producer base of a food chain, therefore, has its greatest effect on the top-level consumer.

Ecosystem populations constantly fluctuate in response to changes in the environment, such as rainfall, mean temperature, and available sunlight.

Normally, such changes are not drastic enough to significantly alter ecosystems, but catastrophic events such as floods, fires and volcanoes can devastate communities and ecosystems. It may be long after such a catastrophic event before a new, mature ecosystem can become established. After severe disturbance the make up of a community is changed. The resulting community of species changes, as early, post disturbance, fast-growing species are out-competed by other species. This natural process is called ecological succession. It involves two types of succession: primary succession and secondary succession.

Primary succession is the development of the first biota in a given region where no life is found. An example is of this is the surrounding areas where volcanic lava has completely covered a region or has built up a new island in the ocean. Initially, only pioneer species can survive there, typically lichens and mosses, which are able to withstand poor conditions.

They are able to survive in highly exposed areas with limited water and nutrients. Lichen, which is made up of both a fungus and an alga, survives by mutualism. The fungus produces an acid, which acts to further dissolve the barren rock. The alga uses those exposed nutrients, along with photosynthesis, to produce food for both. Grass seeds may land in the cracks, carried by wind or birds. The grass grows, further cracking the rocks, and upon completing its own life cycle, contributes organic matter to the crumbling rock to make soil. In time, larger plants, such as shrubs and trees may inhabit the area, offering habitats and niches to immigrating animal life. When the maximum biota that the ecosystem can support is reached, the climax community prevails. This occurs after hundreds if not thousands of years depending on the climate and location.

Secondary succession begins at a different point, when an existing ecosystem's community of species is removed by fire, deforestation, or a bulldozer's work in a vacant lot, leaving only soil. The first few centimeters of this soil may have taken 1000 years to develop from solid rock.

It may be rich in humus, organic waste, and may be stocked with ready seeds of future plants. Secondary succession is also a new beginning, but one with a much quicker regrowth of organisms. Depending on the environment, succession to a climax community may only require 100 to 200 years with normal climate conditions, with communities progressing through stages of early plant and animal species, mid-species and late successional species. Some ecosystems, however, can never by regained.

The grass grows, further cracking the rocks and upon completing its own life cycle, contributes organic content to the crumbling rock becoming soil.

In time, larger plants then shrubs then trees may dominate this area, offering habitats and niches to immigrating animal life. Reaching the maximum biota this ecosystem can support, the climax community caps off further succession after hundreds if not thousands of years depending on the climate and location. Secondary succession begins at a different point. Due to a disturbance to an existing ecosystem, such as fire, deforestation, farmland left abandoned, or even the bulldozer's work in a vacant lot, biota are removed. What does remain, however, is the soil. Soil that may have taken 1000 years to develop the first centimeter from solid rock. Soil that may be rich in humus, organic wastes. Soil that may be stocked with ready seeds of future plants. Secondary succession is also new beginning, but with the advantage of quick regrowth of organisms.

Once underway, the ecosystem's succession to its climax community may only require 100-200 years with normal climate conditions, progressing just like the primary succession through stages of early plant and animal species, mid-species, and late successional species.

Biomes

The biosphere can be divided into relatively large regions called biomes. A biome has a distinct climate and certain living organisms (especially vegetation) characteristic to the region and may contain many ecosystems. The key factors determining climate are average annual precipitation and temperature. These factors, in turn, depend on the geography of the region, such as the latitude and elevation of the region, and mountainous barriers. The major types of biomes include: aquatic, desert, forest, grassland and tundra. Biomes have no distinct boundaries. Instead, there is a transition zone called an ecotone, which contains a variety of plants and animals. For example, an ecotone might be a transition region between a grassland and a desert, with species from both.

Water covers a major portion of the earth's surface, so aquatic biomes contain a rich diversity of plants and animals. Aquatic biomes can be subdivided into two basic types: freshwater and marine. Freshwater has a low salt concentration, usually less than 1 percent, and occurs in several types of regions: ponds and lakes, streams and rivers, and wetlands. Ponds and lakes range in size, and small ponds may be seasonal. They sometimes have limited species diversity due to isolation from other water environments. They can get their water from precipitation, surface runoff, rivers, and springs. Streams and rivers are bodies of flowing water moving in one general direction (i.e., downstream). Streams and rivers start at their upstream headwaters, which could be springs, snowmelt or even lakes.

They continue downstream to their mouths, which may be another stream, river, lake or ocean. The environment of a stream or river may change along its length, ranging from clear, cool water near the head, to warm, sediment-rich water near the mouth. The greatest diversity of living organisms usually occurs in the middle region. Wetlands are places of still water that support aquatic plants, such as cattails, pond lilies and cypress trees. Types of wetlands include marshes, swamps and bogs. Wetlands have the highest diversity of species with many species of birds, fur-bearing mammals, amphibians and reptiles. Some wetlands, such as salt marshes, are not freshwater regions.

Marine regions cover nearly three-fourths of the earth's surface. Marine bodies are salty, having approximately 35 grams of dissolved salt per liter of water (3.5 percent). Oceans are very large marine bodies that dominate the earth's surface and hold the largest ecosystems. They contain a rich diversity of living organisms. Ocean regions can be separated into four major zones: intertidal, pelagic, benthic and abyssal. The intertidal zone is where the ocean meets the land. Sometimes, it is submerged and at other times exposed, depending upon waves and tides. The pelagic zone includes the open ocean further away from land. The benthic zone is the region below the pelagic zone, but not including the very deepest parts of the ocean. The bottom of this zone consists of sediments. The deepest parts of the ocean are known as the abyssal zone. This zone is very cold (near freezing temperatures), and under great pressure from the overlying mass of water. Mid-ocean ridges occur on the ocean floor in abyssal zones. Coral reefs are found in the warm, clear, shallow waters of tropical oceans around islands or along continental coastlines.

They are mostly formed from calcium carbonate produced by living coral. Reefs provide food and shelter for other organisms and protect shorelines from erosion. Estuaries are partially enclosed areas where fresh water and silt from streams or rivers mix with salty ocean water. They represent a transition from land to sea and from freshwater to saltwater. Estuaries are biologically very productive areas and provide homes for a wide variety of plants, birds and animals.

Deserts have relatively little vegetation and the substrate consists mostly of sand, gravel or rocks. The transition regions between deserts and grasslands are sometimes called semiarid deserts (e.g. the Great Basin of the western United States).

Grasslands cover regions where moderate rainfall is sufficient for the growth of grasses, but not enough for stands of trees. There are two main types of grasslands: tropical grasslands (savannas) and temperate grasslands. Tropical grasslands occur in warm climates such as Africa and very limited regions of Australia. They have a few scattered trees and shrubs, but their distinct rainy and dry seasons prevent the formation of tropical forests. Lower rainfall, more variable winter-through-summer temperatures and a near lack of trees characterize temperate grasslands. Prairies are temperate grasslands at fairly high elevation. They may be dominated by long or short grass species.

The vast prairies originally covering central North America, or the Great Plains, were the result of favorable climate conditions created by their high elevation and proximity to the Rocky Mountains. Because temperate grasslands are treeless, relatively flat and have rich soil, most have been replaced by farmland.

Forests are dominated by trees and can be divided into three types: tropical forests, temperate forests and boreal forests. Tropical forests are always warm and wet and are found at lower latitudes. Their annual precipitation is very high, although some regions may have distinct wet and dry seasons. Tropical forests have the highest biodiversity of this biome. Temperate forests occur at mid-latitudes (i.e., North America), and therefore have distinct seasons. Summers are warm and winters are cold. The temperate forests have suffered considerable alteration by humans, who have cleared much of the forest land for fuel, building materials and agricultural use. Boreal forests are located in higher latitudes, like Siberia, where they are known as "taiga."

They have very long, cold winters and a short summer season when most of the precipitation occurs. Boreal forests represent the largest biome on the continents.

Very low temperatures, little precipitation and low biodiversity characterize tundra. Its vegetation is very simple, with virtually no trees. The tundra can be divided into two different types: arctic tundra and alpine tundra. The arctic alpine occurs in polar regions. It has a very short summer growing season. Water collects in ponds and bogs, and the ground has a subsurface layer of permanently frozen soil known as permafrost. Alpine tundra is found at high elevations in tall mountains. The temperatures are not as low as in the arctic tundra, and it has a longer summer growing season.

Evolution of Life

Wherever they are found in the biosphere, living organisms are necessarily linked to their environment. Ecosystems are dynamic and communities change over time in response to abiotic or biotic changes in the environment. For example, the climate may be become warmer or colder, wetter or drier, or the food chain may be disrupted by the loss of a particular population or the introduction of a new one. Species must be able to adapt to these changes in order to survive. As they adapt, the organisms themselves undergo change. Evolution is the gradual change in the genetic makeup of a population of a species over time. It is important to note that it is the population that evolves, rather than individuals.

A species evolves to a particular niche either by adapting to use a niche's environment or adapting to avoid competition with another species. Recall that no two species can occupy the exact same niche in an ecosystem. The availability of resources is pivotal.

In the case of five warbler species which all consume insects of the same tree, to survive each species needs to gather its food (insects) in different parts of that tree. This avoids competition and the possible extinction of one or more species. Therefore, one of the bird species will adapt to hunting at the treetops; another the lowest branches; another the mid-section. In this way, these species have evolved into different, yet similar, niches. All five species in this way can survive by adapting to a narrow niche. Organisms with a narrow niche are called specialized species. Another example is a species that may evolve to a narrow niche by consuming only one type of leaf, such as the Giant Panda, which consumes bamboo leaves.

This strategy allows it to co-exist with another consumer by not competing with it. In both cases, species with a narrow niche are often vulnerable to extinction because they typically cannot respond to changes in the environment. Evolving to a new niche would take too much time for the specialized species under the duress of a drought, for example. On the other hand, a species that can use many foods and locations in which to hunt or gather are known as generalized species. In the event of a drought, a generalized species such as a cockroach may be more successful in finding alternative forms of food, and will survive and reproduce.

Yet another form of evolution is co-evolution, where species adapt to one another by interacting closely. This relationship can be a predator-prey type of interaction. Prey is at risk, but as a species it has evolved chemical defenses or behaviours. On the other hand, co-evolution can be a mutualistic relationship, often characterized by the ants and an acacia tree of South America. The acacia provides ants with food and a habitat, and its large projecting thorns provides protection from predators. The ants, in turn, protect the tree by attacking any animal landing on it and by clearing vegetation at its base. So closely evolved are the species that neither can exist without the other. Similar ecosystems may offer similar niches to organisms, that are adapted or evolved to that niche. Convergent evolution is the development of similar adaptations in two species occupying different yet similar ecosystems. Two species evolve independently to respond to the demands of their ecosystem, and they develop the same mechanism to do so. What emerges are adaptations that resemble look-alikes: Wings of birds and bats are similar, but evolved separately to meet the demands of flying through air. The dolphin, a mammal, shares adaptations that allow for movement through water with the extinct reptile ichthyosaur. They have similar streamlined shapes of fins, head, and nose, which make the bodies better suited for swimming. Natural selection is another process that depends on an organism's ability to survive in a changing environment. While evolution is the gradual change of the genetic makeup over time, natural selection is the force that favors a beneficial set of genes.

For example, birds migrating to an island face competition for the insects on a tropical tree. One genetic pool of a new generation may include a longer beak, which allows the bird to reach into a tropical flower for its nectar. When high populations of birds compete for insects, this ability to use the niche of collecting nectar favours that bird's survival. The long-beaked gene is passed to the next generation and the next, because birds can coexist with the insect-gathering birds by using a different niche. Through reproduction of the surviving longer-beaked birds, natural selection favours its adaptability. A species, family or larger group of organisms may eventually come to the end of its evolutionary line. This is known as extinction. While bad news for those that become extinct, it's a natural occurrence that has been taking place since the beginning of life on earth. Extinctions of species are constantly occurring at some background rate, which is normally matched by speciation. Thus, in the natural world, there is a constant turnover of species.

Occasionally large numbers of species have become extinct over a relatively short geologic time period. The largest mass extinction event in the earth's history occurred at the end of the Permian period, 245 million years ago. As many as 96 percent of all marine species were lost, while on land more than 75 percent of all vertebrate families became extinct. Although, the actual cause of that extinction is unclear, the consensus is that climate change, resulting from sea level change and increased volcanic activity, was an important factor. The most famous of all mass extinctions occurred at the boundary of the Cretaceous and Tertiary periods, 65 million years ago. About 85 percent of species became extinct, including all of the dinosaurs. Most scientists believe that the impact of a small asteroid near the Yucatan Peninsula in Mexico triggered that extinction event. The impact probably induced a dramatic change in the world climate. The most serious extinction of mammals occurred about 11,000 years ago, as the last Ice Age was ending. Over a period of just a few centuries, most of the large mammals around the world, such as the mammoth, became extinct. While climate change may have been a factor in their extinction, a new force had also emerged on the earth - modern humans. Humans, aided by new, sharp-pointed weapons and hunting techniques, may have hurried the demise of the large land mammals.

Over the years, human activity has continued to send many species to an early extinction. The best known examples are the passenger pigeon and the dodo bird, but numerous other species, many of them unknown, are killed off by over harvesting and other human-caused habitat destruction, degradation and fragmentation.

Our biosphere is the global sum of all ecosystems. It can also be called the zone of life on Earth, a closed (apart from solar and cosmic radiation) and self-regulating system. From the broadest biophysiological point of view, the biosphere is the global ecological system integrating all living beings and their relationships, including their interaction with the elements of the lithosphere, hydrosphere and atmosphere.

The biosphere is postulated to have evolved, beginning through a process of biogenesis or biopoesis, at least some 3.5 billion years ago. In a broader sense; biospheres are any closed, self-regulating systems containing ecosystems; including artificial ones such as Biosphere 2 and BIOS-3; and, potentially, ones on other planets or moons.

Origin and Use of the Term

The term "biosphere" was coined by geologist Eduard Suess in 1875, which he defined as:

"The place on Earth's surface where life dwells."

While this concept has a geological origin, it is an indication of the impact of both Darwin and Maury on the earth sciences. The biosphere's ecological context comes from the 1920s, preceding the 1935 introduction of the term "ecosystem" by Sir Arthur Tansley. Vernadsky defined ecology as the science of the biosphere. It is an interdisciplinary concept for integrating astronomy, geophysics, meteorology, biogeography, evolution, geology, geochemistry, hydrology and, generally speaking, all life and earth sciences.

Gaia Hypothesis

The concept that the biosphere is itself a living organism, either actually or metaphorically, is known as the Gaia hypothesis.

James Lovelock, an atmospheric scientist from the United Kingdom, proposed the Gaia hypothesis to explain how biotic and abiotic factors interact in the biosphere. This hypothesis considers Earth itself a kind of living organism. Its atmosphere, geosphere, and hydrosphere are cooperating systems that yield a biosphere full of life. In the early 1970s, Lynn Margulis, a microbiologist from the United States, added to the hypothesis, specifically noting the ties between the biosphere and other Earth systems. For example, when carbon dioxide levels increase in the atmosphere, plants grow more quickly. As their growth continues, they remove more and more carbon dioxide from the atmosphere. Many scientists are now involved in new fields of study that examine interactions between biotic and abiotic factors in the biosphere, such as geobiology and geomicrobiology.

Ecosystems occur when communities and their physical environment work together as a system. The difference between this and a biosphere is simple, the biosphere is everything in general terms.

Extent of Earth's Biosphere

Every part of the planet, from the polar ice caps to the Equator, supports life of some kind. Recent advances in microbiology have demonstrated that microbes live deep beneath the Earth's terrestrial surface, and that the total mass of microbial life in so-called "uninhabitable zones" may, in biomass, exceed all animal and plant life on the surface. The actual thickness of the biosphere on earth is difficult to measure. Birds typically fly at altitudes of 650 to 1,800 meters, and fish that live deep underwater can be found down to -8,372 meters in the Puerto Rico Trench. There are more extreme examples for life on the planet: Rüppell's Vulture has been found at altitudes of 11,300 meters; Bar-headed Geese migrate at altitudes of at least 8,300 meters (over Mount Everest); Yaks live at elevations between 3,200 to 5,400 meters above sea level; mountain goats live up to 3,050 meters. Herbivorous animals at these elevations depend on lichens, grasses, and herbs.

Microscopic organisms live at such extremes that, taking them into consideration puts the thickness of the biosphere much greater. Culturable microbes have been found in the Earth's upper atmosphere as high as 41 km (25 mi) (Wainwright et al., 2003, in FEMS Microbiology Letters).

It is unlikely, however, that microbes are active at such altitudes, where temperatures and air pressure are extremely low and ultraviolet radiation very high. More likely these microbes were brought into the upper atmosphere by winds or possibly volcanic eruptions. Barophilic marine microbes have been found at more than 10 km (6 mi) depth in the Marianas Trench.

Microbes are not limited to the air, water or the Earth's surface. Culturable thermophilic microbes have been extracted from cores drilled more than 5 km (3 mi) into the Earth's crust in Sweden, from rocks between 65-75 °C. Temperature increases with increasing depth into the Earth's crust.

The speed at which the temperature increases depends on many factors, including type of crust (continental vs. oceanic), rock type, geographic location, etc. The upper known limit of microbial is 122 °C (*Methanopyrus kandleri* Strain 116), and it is likely that the limit of life in the "deep biosphere" is defined by temperature rather than absolute depth.

Our biosphere is divided into a number of biomes, inhabited by broadly similar flora and fauna. On land, biomes are separated primarily by latitude.

Terrestrial biomes lying within the Arctic and Antarctic Circles are relatively barren of plant and animal life, while most of the more populous biomes lie near the equator. Terrestrial organisms in temperate and Arctic biomes have relatively small amounts of total biomass, smaller energy budgets, and display prominent adaptations to cold, including world-spanning migrations, social adaptations, homeothermy, estivation and multiple layers of insulation.

BACTERIAL GENOME EVOLUTIONARY DYNAMICS

Gene Loss / Genome Decay

Gene loss or genome decay occurs when a gene is no longer used by the microbe or when a microbe attempts to adapt to a new ecological niche.

Sequencing efforts and microarray analysis have exposed a large number of pseudo genes in some bacterial pathogen species. Mycobacterium leprae for example has been found to contain nearly as many pseudo genes as function genes. M. leprae is not the only microbe exhibiting such behaviour; in his article, Dr. Pallen reports similar properties from Yersinia pestis (the plague pathogen) and also Salmonella enterica. The inactivation of genes is typically associated with a change in the lifestyle of an organism, which can involve adapting to a new niche. The presence of extensive pseudogenes is contrary to another orthodox belief that all the genes in a bacterial genome are functional for some purpose.

It is possible to detect the presence of pseudogenes and the marks of genome decay through whole-genome sequencing in combination with comparative genomics Comparative genomics has helped to reveal that pathogens may favour losing genes in order to live in a host-associated niche and become endosymbionts. Sometimes the shedding of certain genes also renders a pathogen microbe harmless. The analysis of Listeria strains, for example, has shown that a reduced genome size has led to the generation of a non-pathogenic Listeria strain from a pathogenic precursor.

Gene Gain/ Gene Duplication

One of the key forces driving gene gain is thought to be horizontal (lateral) gene transfer (LGT). It is of particular interest in microbial studies because these mobile genetic elements may introduce virulence factors into a new genome. An important comparative study conducted by Gill et al. in 2005 postulated that LGT may have been the cause for pathogen variations between Staphylococcus epidermidis and Staphylococcus aureus. There still, however, remains scepticism about the frequency of LGT, its identification, and its impact.

New and improved methodologies have been engaged, especially in the study of phylogenetics, to validate the presence and effect of LGT.

Gene gain and gene duplication events are balanced by gene loss, such that despite their dynamic nature, the genome of a bacterial species remains approximately the same size.

Genome Rearrangement

Mobile genetic insertion sequences can play a role in genome rearrangement activities. Pathogens that do not live in an isolated environment have been found to contain a large number of insertion sequence elements and various repetitive segments of DNA. The combination of these two genetic elements is thought help mediate homologous recombination. There are pathogens, such as Burkholderia mallei, and Burkholderia pseudomallei which have been shown to exhibit genome-wide rearrangements due to insertion sequences and repetitive DNA segments. At this time, no studies demonstrate genome-wide rearrangement events directly giving rise to pathogenic behaviour in a microbe. This does not mean it is not possible. Genome-wide rearrangements do, however, contribute to the plasticity of bacterial genome, which may prime the conditions for other factors to introduce, or lose, virulence factors.

Single Nucleotide Polymorphism

Single nucleotide polymorphisms(SNPs) are also a genomic variable that adds to the diversity of pathogen strains. Current efforts attempt to catalogue the various SNPs in pathogen strains.

Analysis of Genomic Diversity

There is a need to analyze more than a single genome sequence of a pathogen species to understand pathogen mechanisms. Comparative genomics is a powerful methodology that has gained more applicability with the recent increased amount of sequence information. There are several examples of successful comparative genomics studies, among them the analysis of Listeria. and Escherichia coli. The most important topic comparative genomics, in a pathogenomic context, attempts to address the difference between pathogenic and non-pathogenic microbes. This inquiry, though, proves to be very difficult to analyze, since a single bacterial species can have many strains and the genomic content of each of these strains can vary.

Pan-genomes and Core Genomes

The diversity within pathogen genomes makes it difficult to identify the total number of genes that are associated within all strains of a pathogen species. In fact, it has been thought that the total number of genes associated with a single pathogen species may be unlimited, although some groups are attempting to derive a more empirical value.. For this reason it was necessary to introduce the concept of pan-genomes and core genomes. Pan-genome and core genome literature also tends to have a bias towards reporting on prokaryotic pathogen organisms. Caution may need to be exercised when extending the definition of a pan-genome or a core-genome to the other pathogen organisms; this is because there is no formal evidence of the properties of these pan-genomes. Here, it will be assumed that the definitions may in fact extend, since all pathogen organisms share in the same dynamic genomic events and rely upon variability within strains as a mechanism of survival and virulence.

A core genome is the set of genes found across all strains of a pathogen species. A pan-genome is the entire gene pool for that pathogen species, and includes genes that are not shared by all strains. Pan-genomes may be open or closed depending on whether comparative analysis of multiple strains reveals no new genes (closed) or many new genes (open) compared to the core genome for that pathogen species. In the open pan-genome, genes may be further characterized as dispensable or strain specific. Dispensable genes are those found in more than one strain, but not in all strains, of a pathogen species. Strain specific genes are those found only in one strain of a pathogen species. The differences in pan-genomes are reflections of the life style of the organism. For example, Streptococcus agalactiae, which exists in diverse biological niches, has a broader pan-genome when compared with the more environmentally isolated Bacillus anthracis. Comparative genomics approaches are also being used to understand more about the pan-genome.

Mobile Genetic Elements that Encode Virulence Factors

Three genetic elements of human-affecting pathogens contribute to the transfer of virulence factors: plasmids, pathogenicity island, and prophages. Pathogencity islands and their detection are the focous of several bioinformatics efforts involved in pathogenomics.

Analyzing Microbe-Microbe Interactions

Microbe-host interactions tend to overshadow the consideration of microbe-microbe interactions. Microbe-microbe interactions though can lead to chronic states of infirmity that are difficult to understand and treat.

Bioflims

Biofilms are an example of microbe-microbe interactions and are thought to be associated with up to 80% of human infections. Recently it has been shown that there are specific genes and cell surface proteins involved in the formation of biofilm.These genes and also surface proteins may be characterized through in silico methods to form an expression profile of biofilm interacting bacteria. This expression profile may be used in subsequent analysis of other microbes to predict biofilm microbe behaviour, or to understand how to dismantle biofilm formation.

MOLECULAR BIOLOGY APPROACH TO ENVIRONMENTAL MICROBIOLOGY

Just as the advent of magnifying lenses opened a new area in scientific knowledge and research into the hitherto unknown world of microorganisms, advances in molecular biology and genetic studies are now available to the environmental microbiologist. Our research has employed the use of microbiological culture media and biochemical tests in the identification of microbial populations in environmental samples. Also, different types of microscopic methods have been used in our studies.

The trend now is on molecular approaches based on nucleic acid sequence analysis for direct measure of the abundance, diversity and phylogeny of microorganisms in the environment. This hereditary material (DNA or 16sRNA) is specific to species that can now be identified in a mixed population that is typical of environmental samples. Also the use of nucleic acid probes allows the determination of the taxonomic status of individual cells in complex assemblages in the ecosystem (Hurst, 1997). This method has the advantage that cells can be identified irrespective of their state of health. Thus, viable and nonculturable bacteria in the population can be included in the ecosystem. Moreover, there will then be little use of pure cultures that do not occur in most environmental samples.

Thus during my visit to Rutgers University, New Brunswick, USA in 2000, I worked in collaboration with Douglas Eveleigh using molecular methods to estimate the microbial biomass of an enclosed environment. Samples were taken from the alimentary tract of the giant African land snail, *Archachatina marginata* Swainson. Bacterial DNA in the samples was extracted for sequencing. The ability of gut bacteria to produce enzymes that breakdown components of snail feedstuff was also investigated. This was studied with a view to elucidating the roles of bacteria in digesting plant materials in snail gut ecosystem. Previous studies at the University of Lagos (Odiete & Akpata, 1983) revealed the presence of two undescribed or 'new' bacteria species namely; *Corynebacterium sp nov.* and *Flavobacterium sp nov.* The results are encouraging in the search of unusual environments for microbes with novel capabilities to meet the demands of industries.

ROLES OF AQUATIC MICROORGANISMS IN ENVIRONMENTAL MANAGEMENT

From the discussion it is clear that aquatic microbes impact on man and the environment in various ways and they play important roles in environmental management. Microbes are used for assessing the quality of water. For instance, if the count of heterotrophic bacteria is high it is a sign of organic matter in a water source, hence the water will be unsuitable for culinary purposes.

However there are many uses of water and the quality required depends on the use intended. Microbes are useful in prediction of impacts. The presence of *E. coli* indicates recent contamination of water with sewage or faeces with a high probability of the presence of pathogens in the water. Therefore it can be predicted that consumers of the water may suffer infection of the gastrointestinal tract. Isolation of *Clostridium perfringes* or enterococci indicates that the water had been contaminated at sometime. It is suggested that such water be studied to locate the source of pollution and necessary measures for abatement be taken immediately. In the abatement of pollution, standards are set of microbes that are monitored. Industries and organizations keep within the allowable limits or pay the penalty for non compliance. Treatment of industrial waste waters or domestic sewage is mandatory. A microcosm of microbes brings about the biodegradation of waste matter in the sewage treatment plant.

Unlike raw sewage, treated effluent is more amenable to the receiving environment on disposal. This shows that microbes are used in setting standards and in sewage treatment. Furthermore, the peculiar biodegradative abilities of some microbes have been employed in the rehabilitation of polluted environments. Some *Pseudomonas* species have been exploited in biological remediation of oil polluted environments because the bacteria can breakdown the crude oil.

Conclusion

Attention has been drawn to the fact that evolution of life started in the aquatic environment and microbes were the first formed living cells. To date some microbes remain in their original habitat of water which is abundant on the surface as well as in the aquifer on planet Earth. Microbes isolated from water and especially those in unusual or extreme habitats may have potentials that are novel and useful to man. Water is essential for sustainability of life. Unfortunately, the degradation of water quality is largely by human activities that impede the natural ability of autochthonous microbes to effect self-purification of water. Microorganisms are useful in sewage treatment and in water quality assessment. Knowledge of aquatic microbial ecology may be used in the control of water pollution.

Environmental Impact Assessment (EIA) and maintenance of a healthy aquatic ecosystem would ensure sustainable development.

RECOMMENDATIONS

The responsibilities of a university professor may be grouped into three: Training of high level man power, carrying out research to generate new ideas and knowledge for sustainable development, and participation in the management of academic resources in terms of administration at different levels of the community.

It is customary at the end of an inaugural lecture to send feelers to the public on the relevance of the ivory tower research to national issues. In this lecture some of our research studies were discussed. Under recommendations, I would like to highlight the limitations we face presently in the area of manpower development and also dwell on the need for community involvement in environmental management.

1. Manpower development involves human capacity building. The urriculum for Microbiology at the University of Lagos contains the essential elements required to meet the demands of environmental microbiologists anywhere. However, our students need to be familiar with some newer techniques. They will benefit more from the molecular approach so that they can be more marketable. Unfortunately, it is becoming increasingly difficult to organize enough laboratory practicals to support theory. Even the traditional methods lack space and consumable materials! The University of Lagos was not always like this!! There are too many students for the available spaces, and funds are inadequate for routine laboratory work. The recommendation is that more spaces be provided for practical work so as to enhance our understanding and practice of Microbiology.
2. Environmental Education. The elevation of the Federal Environmental Protection Agency to Federal Ministry of Environment is in line with global trends. Nigeria is playing a leadership role in environmental matters in Africa. Indeed the country is represented at international meetings on the Environment. However, this top administrative posture must be matched with grass-root knowledge and participation in environmental management. Individuals should recognize early signs of environmental degradation and know those to whom a report would receive quick response. Environmental issues are largely what are seen rather than what is said or written! In other words, emphasis should

be on implementation of policies. Our leadership status should be reflected in the level of cleanliness and health of the ecosystem. To achieve this, Environmental Education must be implemented in the curricula of primary and secondary schools. This would infuse environmental awareness into our value system such that environmental consciousness becomes a national culture and a clean environment the pride of all Nigerians.

BIOGEOCHEMICAL CYCLES

While humans cannot control the weather on a daily basis, the influence of human life on the environment plays a significant role in global climate. How often have you wished for a rainy day to go away, or for the warm weather of summer during wintertime? Unfortunately, these wishes rarely come true, and it seems as though humans have little control over the weather. While humans cannot control the weather on a daily basis, the influence of human life on the environment plays a significant role in global climate. Deforestation and fossil fuel burning are just a couple of examples of human activities that seriously disrupt the equilibrium of the global ecosystem and alter the biogeochemical cycles that play a role in determining the Earth's climate. Biogeochemical cycles are essentially the continuous transport and transformation of materials in the environment. Materials are transported through life, air, sea and land in a series of cycles. These cycles include the circulation of elements and nutrients upon which life and the earth's climate depend.

The most important biogeochemical cycles are those of water, carbon, nitrogen and certain other trace gases. In this text, however, we will discuss the carbon and nitrogen cycles, as they are closely intertwined with living things on Earth.

The carbon cycle is particularly influential when it comes to global climate. Much of the carbon in the carbon cycle is in the form of carbon dioxide, a gas that has a strong greenhouse effect because it absorbs infrared radiation.

Carbon is one of the most common elements on Earth and it is the basis of all living things. Below is a graphical depiction of the carbon cycle:

These carbon compounds either:

a) decay into peat, then over millions of years, coal (under very high pressures and worked on by microbes in the absence of oxygen). The coal is then burned by factories to produce electricity, and thus the carbon is returned to carbon dioxide in the air, or
b) are eaten by animals (or remain in the plant, no difference). The carbon compounds in both the plants and animals are returned to the air as

carbon dioxide via respiration and also when they die and decay, as microbes digest their biomass.

BIODEGRADATION OF POLLUTANTS

Microbial biodegradation of pollutants plays a pivotal role in the bioremediation of contaminated soil and ground-water sites. Such pollutants include chloroethenes, steroids, organophosphorus compounds, alkanes, PAHs and PCBs.

Oil Biodegradation

Petroleum oil is toxic, and pollution of the environment by oil causes major ecological concern. Oil spills of coastal regions and the open sea are poorly containable and mitigation is difficult; much of the oil can, however, be eliminated by the hydrocarbon-degrading activities of microbial communities, in particular the hydrocarbonoclastic bacteria (HCB). These organisms can help remedy the ecological damage caused by oil pollution of marine habitats. HCB also have potential biotechnological applications in the areas of bioplastics and biocatalysis.

Waste Biotreatment

Biotreatment, the processing of wastes using living organisms, is an environmentally friendly alternative to other options for treating waste material. Bioreactors have been designed to overcome the various limiting factors of biotreatment processes in highly controlled systems. This versatility in the design of bioreactors allows the treatment of a wide range of wastes under optimized conditions. It is vital to consider various microorganisms and a great number of analyses are often required.

Wastewater Treatment

Wastewater treatment processes are geared towards one purpose: cleaning up water. Recent application of molecular techniques is unveiling the microbial composition and architecture of the complex communities involved in the treatment processes. It is now recognized that wastewater processes harbour a vast variety of microorganisms most of which are yet-to-be cultured, hence uncharacterized. Metagenomic technology is being used to study the diversity, structure and functions of microbial communities in nitrifying processes, anaerobic ammonia oxidation processes and methane fermenting processes.

Environmental Genomics of Cyanobacteria

The application of molecular biology and genomics to environmental microbiology has led to the discovery of a huge complexity in natural communities of microbes. Diversity surveying, community fingerprinting and functional interrogation of natural populations have become common, enabled by a range of molecular and bioinformatics techniques. Recent studies on the ecology of Cyanobacteria have covered many habitats and have demonstrated that cyanobacterial communities tend to be habitat-specific and that much genetic diversity is concealed among morphologically simple types. Molecular, bioinformatics, physiological and geochemical techniques have combined in the study of natural communities of these bacteria.

Corynebacteria

Corynebacteria are a diverse group Gram-positive bacteria found in a range of different ecological niches such as soil, vegetables, sewage, skin, and cheese smear. Some, such as *Corynebacterium diphtheriae*, are important pathogens while others, such as *Corynebacterium glutamicum*, are of immense industrial importance.

C. glutamicum is one of the biotechnologically most important bacterial species with an annual production of more than two million tons of amino acids, mainly L-glutamate and L-lysine.

Legionella

Legionella is common in many environments, with at least 50 species and 70 serogroups identified. *Legionella* is commonly found in aquatic habitats where its ability to survive and to multiply within different protozoa equips the bacterium to be transmissible and pathogenic to humans.

Archaea

Originally, Archaea were once thought of as extremophiles existing only in hostile environments but have since been found in all habitats and may contribute up to 20% of total biomass. Archaea are particularly common in the oceans, and the archaea in plankton may be one of the most abundant groups of organisms on the planet. Archaea are subdivided into four phyla of which two, the Crenarchaeota and the Euryarchaeota, are most intensively studied.

Lactobacillus

Lactobacillus species are found in the environment mainly associated with plant material. They are also found in the gastrointestinal tract of humans, where they are symbiotic and make up a portion of the gut flora.

Aspergillus

Aspergillus spores are common components of aerosols where they drift on air currents, dispersing themselves both short and long distances depending on environmental conditions. When the spores come in contact with a solid or liquid surface, they are deposited and if conditions of moisture are right, they germinate. The ability to disperse globally in air currents and to grow almost anywhere when appropriate food and water are available means that ubiquitous is among the most common adjectives used to describe these moulds.

Microbial Nitrogen Cycling

Microorganisms that convert gaseous nitrogen (N_2) to a form suitable for use by living organisms are pivotal for life on earth. This process is called nitrogen fixation. Another set of microbial reactions utilise the bioavailable nitrogen creating N_2 and completing the cycle in a process called denitrification. This crucial nutrient cycle has long been the subject of extensive research.

Rhizobia

Symbiotic nitrogen fixation is a mutualistic process in which bacteria reside inside plants and reduce atmospheric nitrogen to ammonia. This ammonia can then be used by the plant for the synthesis of proteins and other nitrogen-containing compounds such as nucleic acids. The Gram-negative soil bacteria that carry out this process are collectively referred to as rhizobia (from the Greek words Riza = Root and Bios = Life). The process of symbiotic nitrogen fixation is of agricultural and ecological significance because plants capable of nitrogen fixation do not need to compete for limited quantities of soil nitrogen, nor do they require expensive nitrogenous fertilizers that can be harmful to the environment.

Microalgae

Algae are a highly diverse group of protists, ranging from simple, unicellular organisms to complex, multicellular entities with a range of differentiated tissues and distinct organs. They are found among diverse aquatic ecosystems and play important roles by supplying carbon and energy as well as providing habitat to other members of the biological communities. Some algae cause significant environmental and health problems. There are three algal groups: the dinoflagellates, the diatoms and the haptophytes. Their 3 main phylla are chlorophyta,rhodophyta and phaeophyta.

Anaerobic Protozoa

Diplomonads are a group of mitochondrion-lacking, binucleated flagellates found in anaerobic or micro-aerophilic environments. Most research on diplomonads has focused on *Giardia,* which is a major cause of water-borne enteric disease in humans and other animals. The first diplomonad to have its genome sequenced was a *Giardia* isolate (WB) and the 11.7 million basepair genome is compact in structure and content with simplified basic cellular machineries and metabolism.

Water Microbiology

An adequate supply of safe drinking water is one of the major prerequisites for a healthy life, but waterborne diseases are still a major cause of death in many parts of the world, particularly in young children, the elderly, or those with compromised immune systems. As the epidemiology of waterborne diseases is changing, there is a growing global public health concern about new and reemerging infectious diseases that are occurring through a complex interaction of social, economic, evolutionary, and ecological factors. An important challenge is therefore the rapid, specific and sensitive detection of waterborne pathogens. Presently, microbial tests are based essentially on time-consuming culture methods. However, newer enzymatic, immunological and genetic methods are being developed to replace and/or support classical approaches to microbial detection. Moreover, innovations in nanotechnology and nanosciences are having a significant impact in biodiagnostics, where a number of nanoparticle-based assays and nanodevices have been introduced for biomolecular detection.

Molecular techniques based on genomics, proteomics and transcriptomics are rapidly growing as complete microbial genome sequences are becoming available, and advances are made in sequencing technology, analytical biochemistry, microfluidics and data analysis. While the clinical and food industries are increasingly adapting these techniques, there appear to be major challenges in detecting health-related microbes in source and treated drinking waters. This is due in part to the low density of pathogens in water, necessitating significant processing of large volume samples. From the vast panorama of available molecular techniques, some are finding a place in the water industry: Quantitative PCR, protein detection and immunological approaches, loop-mediated isothermal amplification (LAMP), microarrays.

Sensory Mechanisms

Bacteria have evolved abilities to regulate aspects of their behaviour (such as gene expression) in response to signals in the intracellular and extracellular environment. The interaction of a signal with its receptor (usually a protein or RNA molecule) triggers a series of events that lead to reprogramming of cellular physiology, typically as a consequence of altered patterns of gene expression. In this way, the bacterial cell is able to mount appropriate and effective responses to changing physical and/or chemical environments.

The versatility with which many bacteria adapt to environmental change underlies many important aspects of microbiology. For example, pathogens encounter multiple environments as they invade a host from the outside, and then progress through different sites within host tissues. There is growing evidence that pathogenic bacteria make use of physical and chemical cues to signal their presence in a suitable host, and need to adapt to the host environment in order to mount a successful infection. On the other hand, it should not be assumed that all signals to which bacteria must respond originate in the extracellular environment. For many species, even the cosseted life in a laboratory shake flask is 'stressful', in the sense that there is often a need to avoid or reverse the effects of harmful intermediates or by-products of metabolism. For example, all organisms that use dioxygen as a terminal electron acceptor have to deal with the reactive oxygen species that arise as adventitious by-products of aerobic metabolism. In bacteria, multiple protein receptors for oxygen radicals have been described, which control the expression of genes encoding enzymes that detoxify oxygen radicals or repair the damage that they cause.

Metagenomics

Metagenomics is the cultivation-independent analysis of the collective genomes of microbes within a given environment, using sequence- and function-based approaches. Metagenomic studies have revealed the vast size and richness of the microbial and viral world and demonstrated the phylogenetic diversity of various environments.

Access to huge volume genomic sequence data from uncultured organisms has opened up many new avenues of research. Advances in the throughput of sequencing and screening technologies have greatly facillitated metagenomics research.

EMERGING TECHNOLOGIES: BIOREPORTERS, BIOSENSORS AND MICROPROBES

Bioreporter

Bioreporters are intact, living microbial cells that have been genetically engineered to produce a measurable signal in response to a specific chemical or physical agent in their environment. Bioreporters contain two essential genetic elements, a promoter gene and a reporter gene. The promoter gene is turned on (transcribed) when the target agent is present in the cell's environment. The promoter gene in a normal bacterial cell is linked to other genes that are then likewise transcribed and then translated into proteins that help the cell in either combating or adapting to the agent to which it has been exposed. In the case of a bioreporter, these genes, or portions thereof, have been removed and replaced with a reporter gene. Consequently, turning on the promoter gene now causes the reporter gene to be turned on. Activation of the reporter gene leads to production of reporter proteins that ultimately generate some type of a detectable signal. Therefore, the presence of a signal indicates that the bioreporter has sensed a particular target agent in its environment.

Originally developed for fundamental analysis of factors affecting gene expression, bioreporters were early on applied for the detection of environmental contaminants and have since evolved into fields as diverse as medical diagnostics, precision agriculture, food safety assurance, process monitoring and control, and bio-microelectronic computing. Their versatility stems from the fact that there exist a large number of reporter gene systems that are capable of generating a variety of signals. Additionally, reporter genes can be genetically inserted into bacterial, yeast, plant, and mammalian cells, thereby providing considerable functionality over a wide range of host vectors.

REPORTER GENE SYSTEMS

Several types of reporter genes are available for use in the construction of bioreporter organisms, and the signals they generate can usually be categorized as either colorimetric, fluorescent, luminescent, chemiluminescent or electrochemical. Although each functions differently, their end product always remains the same – a measurable signal that is proportional to the concentration of the unique chemical or physical agent to which they have been exposed. In some instances, the signal only occurs when a secondary substrate is added to the bioassay (*luxAB*, Luc, and aequorin).

For other bioreporters, the signal must be activated by an external light source (GFP and UMT), and for a select few bioreporters, the signal is completely self-induced, with no exogenous substrate or external activation being required (*luxCDABE*). The following sections outline in brief some of the reporter gene systems available and their existing applications.

Bacterial Luciferase (Lux)

Luciferase is a generic name for an enzyme that catalyzes a light-emitting reaction. Luciferases can be found in bacteria, algae, fungi, jellyfish, insects, shrimp, and squid, and the resulting light that these organisms produce is termed bioluminescence. In bacteria, the genes responsible for the light-emitting reaction (the *lux* genes) have been isolated and used extensively in the construction of bioreporters that emit a blue-green light with a maximum intensity at 490 nm. Three variants of *lux* are available, one that functions at

< 30°C, another at < 37°C, and a third at < 45°C. The *lux* genetic system consists of five genes, *luxA, luxB, luxC, luxD,* and *luxE*. Depending on the combination of these genes used, several different types of bioluminescent bioreporters can be constructed.

LuxAB Bioreporters

luxAB bioreporters contain only the *luxA* and *luxB* genes, which together are responsible for generating the light signal. However, to fully complete the light-emitting reaction, a substrate must be supplied to the cell. Typically, this occurs through the addition of the chemical decanal at some point during the bioassay procedure. Numerous *luxAB* bioreporters have been constructed within bacterial, yeast, insect, nematode, plant, and mammalian cell systems.

LuxCDABE Bioreporters

Instead of containing only the *luxA* and *luxB* genes, bioreporters can contain all five genes of the *lux* cassette, thereby allowing for a completely independent light generating system that requires no extraneous additions of substrate nor any excitation by an external light source. So in this bioassay, the bioreporter is simply exposed to a target analyte and a quantitative increase in bioluminescence results, often within less than one hour. Due to their rapidity and ease of use, along with the ability to perform the bioassay repetitively in real-time and on-line, makes *luxCDABE* bioreporters extremely attractive. Consequently, they have been incorporated into a diverse array of detection methodologies ranging from the sensing of environmental contaminants to the real-time monitoring of pathogen infections in living mice.

Nonspecific *lux* Bioreporters

Nonspecific *lux* bioreporters are typically used for the detection of chemical toxins. They are usually designed to continuously bioluminesce. Upon exposure to a chemical toxin, either the cell dies or its metabolic activity is retarded, leading to a decrease in bioluminescent light levels. Their most familiar application is in the Microtox assay where, following a short exposure to several concentrations of the sample, the decreased bioluminescence can be correlated to relative levels of toxicity.

Firefly Luciferase (Luc)

Firefly luciferase catalyzes a reaction that produces visible light in the 550 – 575 nm range. A click-beetle luciferase is also available that produces light at a peak closer to 595 nm. Both luciferases require the addition of an exogenous substrate (luciferin) for the light reaction to occur.

Numerous *luc*-based bioreporters have been constructed for the detection of a wide array of inorganic and organic compounds of environmental concern. Their most promising application, however, probably lies within the field of medical diagnostics. Insertion of the *luc* genes into a human cervical carcinoma cell line (HeLa) illustrated that tumor-cell clearance could be visualized within a living mouse by simply scanning with a charge-coupled device camera, allowing for chemotherapy treatment to rapidly be monitored on-line and in real-time. In another example, the *luc* genes were inserted into human breast cancer cell lines to develop a bioassay for the detection and measurement of substances with potential estrogenic and antiestrogenic activity.

Aequorin

Aequorin is a photoprotein isolated from the bioluminescent jellyfish *Aequorea victoria*. Upon addition of calcium ions (Ca2+) and coelenterazine, a reaction occurs whose end result is the generation of blue light in the 460 - 470 nm range. Aequorin has been incorporated into human B cell lines for the detection of pathogenic bacteria and viruses in what is referred to as the CANARY assay (Cellular Analysis and Notification of Antigen Risks and Yields). The B cells are genetically engineered to produce aequorin. Upon exposure to antigens of different pathogens, the recombinant B cells emit light as a result of activation of an intracellular signaling cascade that releases calcium ions inside the cell.

Green Fluorescent Protein (GFP)

Green fluorescent protein (GFP) is also a photoprotein isolated and cloned from the jellyfish *Aequorea victoria*. Variants have also been isolated from the sea pansy *Renilla reniformis*. GFP, like aequorin, produces a blue fluorescent signal, but without the required addition of an exogenous substrate. All that is required is an ultraviolet light source to activate the fluorescent properties of the photoprotein. This ability to autofluoresce makes GFP highly desirable in biosensing assays since it can be used on-line and in real-time to monitor intact, living cells. Additionally, the ability to alter GFP to produce light emissions besides blue (i.e., cyan, red, and yellow) allows it to be used as a multianalyte detector. Consequently, GFP has been used extensively in bioreporter constructs within bacterial, yeast, nematode, plant, and mammalian hosts.

Uroporphyrinogen (Urogen) III Methyltransferase (UMT)

Uroporphyrinogen (urogen) III methyltransferase (UMT) catalyzes a reaction that yields two fluorescent products which produce a red-orange fluorescence in the 590 - 770 nm range when illuminated with ultraviolet light.

So as with GFP, no addition of exogenous substrates is required. UMT has been used as a bioreporter for the selection of recombinant plasmids, as a marker for gene transcription in bacterial, yeast, and mammalian cells, and for the detection of toxic salts such as arsenite and antimonite.

BIOSENSOR

A biosensor is an analytical device for the detection of an analyte that combines a biological component with a physicochemical detector component.

It consists of 3 parts:

- the *sensitive biological element* (biological material (e.g. tissue, microorganisms, organelles, cell receptors, enzymes, antibodies, nucleic acids, etc.), a biologically derived material or biomimic) The sensitive elements can be created by biological engineering.
- the *transducer* or the *detector element* (works in a physicochemical way; optical, piezoelectric, electrochemical, etc.) that transforms the signal resulting from the interaction of the analyte with the biological element into another signal (i.e., transducers) that can be more easily measured and quantified;
- associated electronics or signal processors that are primarily responsible for the display of the results in a user-friendly way.. This sometimes accounts for the most expensive part of the sensor device, however it is possible to generate a user friendly display that includes transducer and sensitive element.

A common example of a commercial biosensor is the blood glucose biosensor, which uses the enzyme glucose oxidase to break blood glucose down. In doing so it first oxidizes glucose and uses two electrons to reduce the FAD (a component of the enzyme) to FADH2.

This in turn is oxidized by the electrode (accepting two electrons from the electrode) in a number of steps. The resulting current is a measure of the concentration of glucose. In this case, the electrode is the transducer and the enzyme is the biologically active component.

Recently, arrays of many different detector molecules have been applied in so called electronic nose devices, where the pattern of response from the detectors is used to fingerprint a substance.. In the Wasp Hound odor-detector, the mechanical element is a video camera and the biological element is five parasitic wasps who have been conditioned to swarm in response to the presence of a specific chemical. Current commercial electronic noses, however, do not use biological elements.

A canary in a cage, as used by miners to warn of gas, could be considered a biosensor. Many of today's biosensor applications are similar, in that they use organisms which respond to toxic substances at a much lower concentrations than humans can detect to warn of the presence of the toxin. Such devices can be used in environmental monitoring, trace gas detection and in water treatment facilities.

AQUATIC MICROBES SHAPE OUR WORLD

The Earth is a complex and ever-changing combination of land, air, water and living organisms. These components are intricately linked so that processes occurring in one affect the others. Microbes play vital roles in this global system by driving biogeochemical cycles that affect the climate and move elements around the planet, making them available to other organisms. Without microbes, the Earth would be uninhabitable and yet they are the least understood life-forms on our planet.

The Marine and Freshwater Microbial Biodiversity programme has investigated how aquatic microbes shape the global system. Marine microbes play a critical role in regulating natural and man-made gases that affect climate. New research has let us cultivate previously-unknown bacteria that break down methybromide - a chemical that destroys stratospheric ozone. Scientists have investigated microbial methane production, and how marine viruses affect levels of dimethyl sulphide, a gas that promotes cloud formation. Other groups have studied how bacteria fix gaseous nitrogen for growth in seawater, and how they interact with trace metals in freshwater sediments, contaminated estuaries and coastal seas. Methyl bromide gas destroys a massive 25% of ozone lost from the stratosphere. It is released when biomass burns, when cars use leaded petrol, and some countries use it to fumigate soils and harvested crops. Researchers at the University of Warwick and the Plymouth Marine Laboratory have cultivated three new types of bacteria with enzymes that break down the gas. Using molecular probes, they discovered that marine environments harbour many different bacteria with this enzyme.

Methane is increasing in the atmosphere faster than carbon dioxide, and has 21 times its global warming potential. Methane is produced by archaea in the absence of oxygen, and researchers usually assume most of it comes from freshwater environments like wetlands and marshes. Scientists from Cardiff University used tracer and stable isotope techniques to study methane production at 21 sites, ranging from full freshwater to seawater. The team's results will help identify methane sources – a critical step towards reducing inputs to the atmosphere.

Marine micro-organisms also produce beneficial gases, for example dimethyl sulphide (DMS), the dominant sulphur trace gas in seawater. In laboratory culture and field mesocosm studies, researchers at the Marine Biological

Association and the University of East Anglia showed that when viruses kill the marine phytoplankton *Emiliania huxleyi*, DMS production increases. DMS emitted from the sea is important in the global sulphur cycle. It also influences global climate by producing sulphate particles that can promote cloud formation, and hence altering how much of the Sun's energy the Earth reflects back to space.

Microbes can convert nitrogen gas to ammonia and organic nitrogen compounds – a critical process in the global nitrogen cycle. Levels of available nitrogen limit phytoplankton growth in the sea, yet more phytoplankton grow in some tropical and sub-tropical regions than known nitrogen inputs can support. Researchers from the Universities of Stirling and Newcastle used molecular methods to show that as-yet-uncultured nitrogen-fixing bacteria are widely distributed. Scientists used to think that microrganisms only fix nitrogen when they have to, because the process uses a lot of energy and other resources such as iron. So it was surprising to discover that key nitrogen-fixation genes in these bacteria were still operating when ammonia and nitrate were available in the seawater.

Living organisms need trace metals for various processes, but elevated levels can be toxic. Researchers from the Plymouth Marine Laboratory and the University of Southampton looked at how copper and zinc affected the bacterial assemblage, and the size and shape of aggregates (decaying phytoplankton debris and bacteria) in water from clean and metal-contaminated estuaries. They also used micro-electrodes to map oxygen concentration in marine aggregates. Scientists from the University of Lancaster developed a 2D imaging technique to assess how microbes influence geochemical processes in sediments. These projects help show how bacterial activity affects the mobility of trace metals at micro scales.

5

Design and Analysis of Toxicological Environment

Toxicologic studies are either made on animals or on human subjects; the latter being under various observational setups rather than in strict experimental ones. Typically, such studies are made first on subhuman primates to set appropriate dose and exposure levels as well as examine possible side-effects or other forms of toxicity (besides the main end-point under consideration).

These animal studies are generally referred to as *dosimetric assays*. Even such dosimetric studies differ considerably from conventional laboratory experiments which can be usually conducted under relatively more controlled experimental setups. Toxicity in environment and in other natural forms often have latent effects, usually working in a rather synergism mode, and often, may have difficulties in identifying the principal end-point(s).

Because of these reasons, conventional designs that are used in agronomic or industrial studies or even in clinical/ medical studies are often inappropriate for such toxicologic investigations. Although *longitudinal studies* or *repeated measurement designs* have more relevance to toxicologic investigations, simple statistical tools are usually not so appropriate in such contexts due to model complexities, considerable degree of spatial and temporal variations, and the very nature of the response pattern.

We may conceive of a simple dose-response model wherein a simple linear or specific nonlinear regression function with distributional homogeneity of the residuals from the regression is conceived, and often even the error distribution is taken for granted to be Gaussian. This stationarity of the errors in a statistical sense is a fundamental assumption of statistical modelling and analysis schemes, and often blocked designs are used to reduce further the margin of errors.

In a multivariate setup multivariate analysis of variance (MANOVA) models have been developed even under more complex designs relating to multi-factor and multi-response experiments. Here also, distributional homogeneity of the error vectors, additivity of the treatment and factor effects, and normality are taken for granted. In recent years, more emphasis has been laid down on robustness aspects that introduce flexibility to the models.

However, additivity and some other homogeneity conditions remain as vital components of the basic regularity assumptions. In toxicologic studies, especially relating to environmental toxicants, the degree of homogeneity of the distribution of errors, the additivity of the effects, Gaussian distributional patterns may all be questionable to a larger extent. As such, whatever optimality considerations pertaining to standard statistical designs may no longer be pertinent to toxicologic studies. This serious aspect deserves a careful appraisal, and will be made later.

Sampling Schemes

Accurate determination of prevailing toxicant levels and their interactive mode of action need a careful statistical appraisal. In agronomic studies, laboratory experimentation, or in many other fields of application, simple random sampling (SRS), with or without replacement, is generally adopted.

This permits standard statistical tools to draw statistical conclusions subject to assertable levels of margin of errors. In more complex studies, instead of SRS, one may have unequal probability sampling (UPS) schemes, again with or without replacement. In toxicologic studies, particularly in environmental toxicology, the prevalence of toxic elements may not follow a regular or even smooth pattern. Moreover, for all practical purposes, it might not be feasible to record the prevalence data cloud on a continuous basis over a spatial and temporal domain. Usually, a certain number of grid-points are chosen as monitoring sites, and data collected from such points, spanning a time period, are incorporated in the study of spatio-temporal variation over the area spanned by the chosen grid-points. In view of the paucity of such grid-points, enormous (and uncontrolled) variability in the level of contaminants, and rather complex PDPK undercurrents, it is quite important to pay adequate attention to the sampling scheme and the number of observations that may be needed to draw statistical conclusions with some postulated margin of error.

Further, in atmospheric toxicity, usually a three dimensional grid set is used to study the effect of altitude and the distance from a contaminating source, and in addition, nonstandard recording device may be needed. This may complicate the sampling scheme. In many cases a systematic sampling scheme is used for administrative convenience, though that might lead to inappropriateness of standard statistical methodology for drawing objective conclusions from acquired data sets.

Epidemiologic Undercurrents

In a conventional setup, observational studies are usually made for toxicologic investigations involving human beings, while laboratory experiments are advocated for animal studies. Even for such dosimetric studies, a complete randomization may not be always adoptable, and intracluster dependence pattern may cause serious problems in statistical modelling and analysis.

For epidemiologic studies, often, case-control, matching, and cohort analysis methodologies are recommended for better control of bias and undesirable variation. For purely observational studies, there may be a basic issue: how far statistical modelling and analysis can be incorporated in such studies?

In view of the paramount importance of such scientific findings and their impact on our community as well as public health, there is a genuine need to explore appropriate and valid statistical modelling and analysis schemes that would match the type of observational studies employed; we refer to a recent volume where such statistical issues have been addressed thoroughly (by a host of contributing authors). At present, there is considerable emphasis on genetic epidemiology and environmental epidemiology which have also good relevance to toxicology.

Biostatistics and epidemiology have been working in close cooperation in resolving some of the basic issues in such epidemiologic studies. One of the most significant developments in this context is the evolution of clinical epidemiology that borrows strength from pilot dosimetric studies and applies them to human subjects through controlled clinical trials. In this manner, there has been a harmonious blending of epidemiologic, clinical and statistical reasonings.

Biological Assays or Bioassays

These are akin to dosimetric studies; traditionally, a new and a standard preparation (that could be toxic in some sense) are compared by means of reactions that follow their application to some subjects or living organisms, and the basic objective is to study the *relative potency* of the new preparation with respect to the standard one. In this sense, bioassays are somewhat different from toxicologic studies, where there could be a lack of a *control* or *placebo* group against which the *treatment* group should be statistically compared.

Rather, in toxicologic studies, there is a more dominant dose–response relationship emphasis in which the dose levels may not be determined that accurately (nor the responses), and there could be more unaccountable variation due to the complexities of environmental and behavioural factors. Nevertheless, in terms of statistical methodology, bioassays in a broad sense have some common features with toxicologic assays. Radioimmunoassays are good examples in this context. Bioavailability and bioequivalence trials are also used for toxicologic studies.

Whither dose-Response Regression?

Although there is a predominant dose-response flavour in the sense that the dose refers to the various toxic substances and their prevailing levels while the response relates to the reactions that follow the intake of toxic material by various organisms, most notably, human beings, there are some crucial steps that need to be examined before attempting to use standard dose–response regression models. Let us iterate some of these roadblocks and explain the surrounding complexities and difficulties in the light of statistical perspectives.

- *Accurate or reliable identification of the actual toxins and their effective dose levels*: In general, there may be a (large) number of toxic elements (some of which may even be latent) with rather diverse mode of propagation, and hence, putting such a complex dose system may need considerable appraisal from various aspects. From a statistical standpoint, there may be the basic question of identifiability of all or even the majority of toxins that are particularly relevant in a specific context. In a majority of cases, even a simple model with a definite pattern of synergism of these toxins may not totally meet the light of reality of the actual pattern, and hence, elimination of redundant statistical information is a basic problem in such modelling and analysis schemes.
- *Measurement errors and compliance issues*: A basic requirement in a dose-response regression analysis is the precise determination of the dose levels and the responses, no matter whether they are quantitative or quantal. However, in toxicologic studies, even in simple dosimetric ones, an accurate determination of the dose levels may not always be possible. In conventional linear models, measurement errors have been accounted in various forms, and it might be quite intuitive to incorporate some of these findings in the current context. There are some limitations for such standard adoptions. In dosimetric studies and more noticeably in actual environmental toxicology studies, an administered or recorded dose level may be grossly inappropriate as a regressor, and because of differential exposure levels, stochastic compliance error models are more appropriate. Basically, such models are closely based on either Bayesian methodology or historical control and additional information on the compliance pattern.

Biomechanistic and PBPK Modeling

Advances in biotechnology and modern computer technology have made it possible to advocate suitable models that incorporate the basic biological and toxicologic factors in the depiction of the intake process of toxicants, their pathology and biological/toxicologic impact in terms of appropriate biomathematical formulations, and on top of that, impose stochastics to reflect the unaccounted variations associated with the process.

These constitute the *physiologically-based pharmacokinetic* (PBPK) models. In the presence of genotoxicity, such models may generally become more complex due to linkage and other genetic stochastics factors that involve more dominant stochastic elements. For example, growth of a tumour possibly leading to cancer can be related to the intake of certain specific toxicants in certain specific mode, and such stochastic biomechanistic models may be quite useful, particularly for studies involving rodents or other subhuman primates.

In lung diseases, such a biomechanistic model may become more complex due to genetic effects and highly variable environmental effects. Occupational hazards also contribute more toward a complex stochastic environment for which simple models may not work out properly. For this reason, PBPK modelling is often advocated to incorporate more biological and toxicologic understanding of the whole intake process and its aftermath.

At present, there is a lot of emphasis on the so-called *stochastic partial differential equations* (SPDE) which have the natural ability to incorporate stochastics in nice diffusion process formulations, and at the same time, the deterministic factors in terms of the parameters that appear in such SPDEs. On the other hand, in a majority of cases, especially involving human subjects, such SPDEs are quite complex and involve a large number of parameters; statistical interpretations and drawing conclusion from such SPDE models may often become difficult. There is ample room for more developments in this much needed area of statistical research.

Dosimetric Studies

Animal studies or dosimetry can be regarded as the precursors of serious scientific toxicologic studies. Basically, most of the toxicity studies, though aimed for human subjects, cannot be directly applied to them; rather, there is a need for delicate dose and therapeutic evaluations, as well as safety considerations. As a preliminary step to gather such pertinent information, animal studies are usually prescribed, though often with some limitations. There are several advantages of animal studies compared to human studies.

Such studies can be conducted under relatively more controlled laboratory setups, and hence, statistical modelling and analysis aspects are relatively simpler. Also, cost-wise dosimetric studies are more economic than human ones. Moreover, from therapeutic and dose setting points of view, such studies can provide valuable insights for subsequent studies involving human beings.

On the other hand, subhuman primates may vary considerably from human beings in body metabolisms, pharmaceutical dynamics and kinetics, and hence, incorporating the findings from dosimetric studies for designing and modelling of toxicologic studies involving human beings may require a very careful appraisal.

Conventional Regression Models

One of the basic approaches to the dose response regression modelling is to incorporate a suitable linear regression model, following transformation on the dose and/or response variables as needed to induce more linearity of the regression function and/or normality of the error components. In the same vein, generalized linear models14 have been advocated for quantal response models and other types of nonstandard regression forms. In the context of multiresponse models with possibly non-continuous variables and latent effects, the appropriateness of (generalized or general) linear models may often be questionable. In fact, their primary shortcoming may be the lack of robustness property (to plausible departures from modelbased assumptions, even to a moderate extent).

On top of that, measurement or compliance error can introduce considerable complications in such standard models, and may even invalidate the models. For this reason, nonparametrics have been advocated for a wide class of models that arise in dosimetric studies, bioassays and bioequivalence trials which are potentially related to toxicologic studies. Model misspecification and measurement errors in covariates have been studied in this context. Basically, these are all large sample procedures, and there remains considerable scope for developing moderate to finite sample solutions.

Spatio-Temporal Models

Generally, in environmental settings, data are collected on a set of grid-points scattered in a region, and a set of time points covering a period. The matrix of observations is then incorporated in a spatio-temporal setup to draw conclusions for the region and time period. Since, we may have generally count variables, or even qualitative data models, without a reasonably stationary pattern, the usual *variogram*-model based statistical analysis may not be suitable; in fact, the very linearity or additivity of the accountable factors might be questionable, and hence, the usual methods based on *kriging* and spatial homogeneity may not be that appropriate.

Actually, some local homogeneity of spatial covariation is often assumed in order that standard kriging and variogram models can be adapted, though with a slower rate of convergence. Bayesian methodology, under appropriate priors on the spatial variation, is also being applied in this context.

Small area estimation procedures are also being incorporated. A complete nonparametric approach may usually require larger sample sizes.

In this respect, nonparametric and semiparametric longitudinal and spatial data analysis models appear to be more appropriate. This is currently an active area of statistical research, and there is ample room for further developments.

ENVIRONMENTAL TOXICITY

Environmental toxicology is a field of environmental science that studies how exposure to toxic chemicals can directly change the health of individuals or indirectly affect us by disrupting or defiling the content of lakes, oceans, forests, and the atmosphere. Environmental toxicologists also assess the potential risks of substances that may be deemed harmful to people, animals, or other aspects of the environment.

The Air We Breathe

As an interdisciplinary science studying every aspect of the biosphere, environmental toxicology has focused on the effect of man-made pollutants on our atmosphere for the past 40 years. Since discovering that rates of skin cancer were rising along with evidence that the ozone layer was deteriorating, environmental toxicologists have kept detailed statistics on the increasing amount of pollution in the atmosphere and its detrimental effects on human and animal life. Environmental toxicologists cite evidence of rampant pollution in the erosion caused by acid rain.

Aquatic Systems and Excess Nutrients

Although we're taught we should make sure we get plenty of nutrients in our daily diet, environmental toxicology has found evidence that an excess amount of nutrients in runoff from agricultural fields can devastate an aquatic system. Although phosphorous, carbon and nitrogen are essential to proper growth, too much of them can cause a harmful surplus of plant and microbial growth that can choke and kill ecological systems. This causes eutrophication in lakes and streams, which is usually the reason behind massive fish kills. Environmental toxicologists investigate these kinds of habitat disruptions and advise on how to resolve them.

Heavy Metal

Since the industrial revolution of more than 150 years ago, consumer and factory waste products, which erode soils and release heavy metals into our natural water supply, have rapidly contaminated our drinking water. Environmental toxicology researches the amount and effects of these metallic elements that are considered high density and toxic to humans in minimal doses. Some of the more common heavy metals are mercury, arsenic, lead, and chromium, which can accumulate in our bodies. Creating substances that can counteract the effects of these metals is another facet of environmental toxicology.

DNA Mutations

Environmental toxicology encompasses a broad area of the sciences and sometimes overlaps with biotechnology. Research in how environmental toxins can affect DNA structure and produce mutations in both human and animal organisms is another area environmental toxicologists cover. In addition, they also examine detailed aspects regarding the intake, metabolism, and excretion of toxic elements in the air or water. An example of this type of research involves the detrimental consequences on human health resulting from the radioactive fallout following the Chernobyl nuclear reactor disaster.

The Big C – Cancer

A major area of research in the field of environmental toxicology involves how toxic elements released into the environment precipitate the emergence of cancers such as leukemia, lung cancer, lymphomas, and melanoma. An environmental toxicologist will investigate the incidences of such cancers and how intensely exposed an individual or groups of people suffering from cancer have been to certain chemicals. Some of these harmful chemicals include pesticides, asbestos, solvents, and herbicides. In addition, environmental toxicologists are often activists who try to lobby for regulations on these chemicals.

The creation of chemicals meant to facilitate our lives sometimes backfires. At these times, the field of environmental toxicology steps in to determine these chemicals' effects, what we can expect if we continue to use them, and how we might reduce the harmful effects of such damaging chemicals on ourselves and on our environment.

Environmental Smoking

Smoking of cigarette, cigars, bidis and allied tobacco products leads to significant toxicants, not only for the active smokers but also for passive smokers (that is, those who are in the vicinity of active smokers); the mode of intake is primarily inhalation of toxic fumes and carbon monoxide-rich gases. Emission from factories that use coal or wood as fuel, as well as home cooking on coal energy also contribute significantly to environmental smoking.

It is a belief that cigarette smoking is linked to lung cancer and other respiratory diseases, though the cancer etiology is still not that precisely known. The Surgeon General in USA has a statutory label on every cigarette packet: Smoking is dangerous for our health. Are we aware that industrial smoking (for example, fumes from factory exhausts) may be equally, if not more, harmful to our health?

Automobile and Aviation Exhausts

A majority of automobiles (including cars, vans and trucks), and locomotives (including trains) run on diesel and gasoline fuels; so do the aeroplanes of various types (which generally require a high-octane fuel product). Only in some countries, alcohol made from sugarcanes is used as fuel in cars. These conventional fuels contain various toxicants.

On top of that in severe winter time, some other reagents are added to gasoline products to facilitate smooth starting in sub-zero temperature; these are also suspected to have carcinogens and other toxicants. Elimination of lead from gasoline products (on a wider scale) has resulted in some improvements in recent past. Airplane exhausts, while on air, contribute toxicants in the upper atmosphere that with rain and smog try to come down towards the surface level.

OCCUPATIONAL ENVIRONMENT AND TOXICITY

Many of the present time occupational environments are contaminated with toxicants and other health hazards. As examples, we may cite the following:

Coal miners (black-lung) disease: In mining operations all over the world, and more so in third world countries, in the underground tunnels, there may be an excessive amount of airborne carbon particulate materials (APM), carbon monoxide, and other harmful gases. In that way, inhalation toxicology is most pertinent to the coal miners occupational environment, and lung diseases used to be prevalent among them. Subsoil water contamination and moisture, coupled with possibly inadequate air circulation system may add more distress to this toxic environment.

Asbestosis: Asbestos is an incombustible, fibrous mineral that used to be commonly used as an insulation material, and often as a substitute for tin or other metals for corrugated roofing. The white dust particles in an asbestos plant are quite catalytic for respiratory diseases as they may get absorbed in the lung tissues. Again, inhalation mode is the more common intake process, and asbestos has also been identified as a carcinogen.

Even in houses or schools and work places where asbestos used to be the insulation material, the dust particles play the same role, but possibly to a lesser intensive level. Young children often unknowingly placing their hands on such insulations and without washing taking food may induce the ingestion mode, while epidermal contact may have absorption intake mode.

Chemical Plants

DDT is a classic example of such a highly toxic material. Though primarily meant for controlling insects and bugs, it has a significant carcinogenic effect on human beings as well. The list is by no means very tenuous. The 1984 highly toxic gas leakage from the Bhopal Union Carbide plant had such a high toll in terms of mortality and morbidity that the deleterious effects of such toxic gases are evident.

Nuclear power plants: No matter how much precaution is adopted in dumping the nuclear waste products or releasing the (low level) radioactive gases in the atmosphere, such radiation effects can affect human being in inhalation, absorption and ingestion modes. The casualty from various power plant leakage bears this testimony. The impact of nuclear warheads and weapons (including bombs) can be even more disastrous. There are literally thousands of occupational hazards and toxicity sources, and it might be better to check occupational safety and security standards to minimize their impact.

Waste products: Industrial and domestic waste products generally produce a lot of toxic effects. Sulphur PMs, nitrogen dioxide, mercury, carbon monoxide from oil refineries as well as thermal power plants, nuclear and chemical plants waste products, and even the garbage collected from households contain significant toxic materials. All modes of intake are active in this respect.

neuro-toxicology

Our engulfing toxic environment often works silently in slow progression at an early phase of intervention, and then at an opportune time, switches its gear and pounds into deleterious impacts. The intimate relationship between body and mind in the entire animal world (including human being) has been well established. A healthy body (especially at the developmental stage) tends to have harmonious impact on a healthy mind.

Continual health problems, particularly relating to chronic diseases and irreversible disorders, may significantly affect the quality of life and behavioural patterns. As much as malnutrition has a significant influence on mental healthrelated illness, toxicity has a similar impact on personality, outlook on life, and quality of life; this branch of toxicology is referred to as the behavioural toxicology.

Domestic violence, irrational thinking processes (including committing suicide), clinical depression, addiction to drugs and alcoholic drinks may all be linked to the broad spectrum of toxicity that surrounds us in our everyday living. In fact, the aging process is also tied-down to the environmental toxicity in a broad sense, though a direct causal relationship may be hard to establish; genetic toxicology is coming up with good evidence of such behavioural dysfunctions due to toxicity in our bioenvironment.

Epidemiology has long been involved in this field of research through mostly observational or case-control studies. Environmental and genetic epidemiology are also going through evolutionaly growth in acquiring fruitful information in this respect.

Reproductive (and developmental) toxicology deal with adverse effects on male and female resulting from exposures to specific toxicants that may lead to reduced fertility, alterations in sexual behaviour, adverse pregnancy outcomes and alterations of other functions relating to the reproductive system (including depression). On the other hand, developmental toxicity relates to adverse effects in the developing organism due to exposures to either parent before the actual conception or during the prenatal or the postnatal development; these include still-birth, birth defects, teratogenesis and structural abnormality, altered growth, and abnormal functions.

For people exposed regularly to anaesthetic gases, a decrease in fertility and premature birth problems have been observed; among the smokers in their pregnancy, early pregnancy loss, reduced body weight for babies and other birth defects are more commonly observed. Toxic gases, such as dioxin, may have adverse reproductive and developmental toxicity effects.

Neurotoxicology deals with the impact of toxicity (mostly, prevailing in surrounding environment) on human nervous system which is controlled by the central nervous system (CNS) located in the proximity of the brain. Toxic elements in the cardiovascular system resulting from either inhalation of toxic gases or PMs, or ingestion of toxic foods or drinks, and also absorption of toxicity through skin, may generally impair the normal working environment of the CNS.

Reduced flow of oxygen to the cortex area is a significant factor in this respect, and compounded with other toxic elements in the blood in the upper aorta, it can have even far more devastating effects. A less-than-ideal level of functioning of the CNS and the cortex can also affect the personality and sense of security. Thus, behavioural toxicity and neurotoxicity are quite interrelated. Even reproductive toxicity has a good bearing on both behavioural and neurotoxicity.

Statistical perspectives

Outline of toxicology in relation to our body, mind, our bioecosystem, and society as a whole, the basic concern is to identify the toxicity undercurrents, to have a complete inventory of toxins of various kinds, to arouse our awareness to the grave impact of such toxicants, to study their mode of progression and distribution, their (molecular) biological interaction and impact, to quantify the risk due to toxicants, and effectively accomplish the risk assessment task with a view to promoting global safety standards and improving the quality of life.

In all these aspects, statistical reasoning plays a fundamental role. Several broad aspects of statistical perspectives need to be appraised in this setup.

Broadly speaking, collection of relevant toxicologic data that can be put to statistical modelling and analysis schemes needs careful planning as well as skillful sampling techniques (as a complete census is often ruled out). In many situations, we may have, at best, some observational studies (mostly posed from epidemiologic considerations), and it could be a challenging task to validate statistical analysis in order to procure objective conclusions. For this reason, epidemiology and biostatistics need to be blended to accomplish this task as effectively as possible.

Animal or dosimetric studies are often advocated to gather information of therapeutic and toxic factors and their impact on the end-point(s) in mind. There are, however, some roadblocks to such dosimetric studies, especially in environmental toxicology, and hence, we need to appraise them too. Bioassays are more often used in pharmacologic setups, and are quite useful in toxicologic studies as well. There are some other statistical issues relating to toxicology that deserve some appraisal too.

SEDIMENTS TOXICOLOGY

Sediments can act as a sink for contaminants entering the aquatic ecosystem, or they may serve as a source of contamination. These toxins can accumulate in benthic organisms and alter the health of these organisms, act as an entry point for contaminants into the food web, and reduce the ecological integrity of the aquatic environment.

Physiochemical differences in sediment and the ecology of benthic organisms influence contaminant exposure and uptake. The multifaceted nature of sediment toxicity requires wildlife and landscape scientists to work with the larger Environment Canada community to develop a variety of approaches required to determine the dynamic characteristics of contaminated sediment, and the influence on resident biota.

Contaminated sediments are a global environmental issue known to affect North America, Europe, and Australasia. This is because sediments serve as an efficient sink for many anthropogenic contaminants associated with various types of industrial activity, including nonionic organic chemicals (NOCs), such as polychlorinated biphenyls (PCBs) and polycyclic aromatic hydrocarbons (PAHs), and cationic metals such as cadmium, copper, nickel, and lead.

For example, in the United States, the U.S. Environmental Protection Agency's Contaminated Sediment Inventory and the National Oceanic and Atmospheric Administration Status and Trends Program have reported large areas of sediments contaminated with anthropogenic chemicals. The affected areas have been quantified to represent billions of metric tons of contaminated sediment. The presence of these contaminants can have several adverse effects to the environment, ranging from acute and sublethal toxicity to invertebrates and fish to benthic community impairments.

The implications of these effects are both ecological and economic in nature. For example, severely toxic sediments could significantly impair a benthic invertebrate community fed upon by a commercially important fish. In turn, the lack of a benthic community could result in a nonviable fishery.

The principle approach to clean up such contaminated sediments is through remediation. Currently, contaminated sediment remediation methods include environmental dredging, in which the contaminated sediments are removed from a site, monitored natural recovery, in which the contaminated sediments of the site are allowed to be buried naturally through siltation with clean sediments, and in situ capping. Capping is similar to monitored natural recovery, but it is enhanced by placing material on top of the contaminated site.

In many instances, capping material is simply sand; however, other more chemically active capping materials have been investigated, including the addition of various forms of activated carbon. The addition of active capping materials seeks to enhance the sequestering of contaminants within the cap and reduce the transport and bioavailability of toxic chemicals associated with the contaminated sediments.

Recently, the use of activated carbon as a sediment amendment and potential active capping material has been explored for PCBs, PAHs, and pesticides. These evaluations have also shown promise for possible field application. For example, for sediments from the Lake Hartwell Superfund site, Werner et al. reported reducing PCB aqueous concentrations by up to 95 per cent and accumulation by semipermeable membrane devices by 78 per cent after one month of activated carbon–sediment interaction.

Furthermore, in a preliminary field study at Hunter's Point Shipyard, Cho et al.demonstrated that activated carbon could be effectively distributed into contaminated sediments and result in evidence of reduced PCB bioavailability to the bivalve *Macoma nasuta* several months after activated carbon deployment.

Coal fly ash, another material that has a long history of use in pollution control, is now being considered for capping. Coal fly ash is most often formed during the combustion of coal in the production of electricity at power plants. Like activated carbon, fly ash is formed in a process that can produce a type of refractory and highly sorptive black carbon. In recent studies, black carbon has been shown to strongly adsorb many NOCs.

Several studies have examined the interaction of NOCs and cationic metals with fly ash, demonstrating strong adsorption under some conditions. Fly ash is an appealing remedial material because it is inexpensive, readily available from the very large quantities produced annually from the burning of coal, and reuses a waste substance. For example, according to the American Coal Ash Association and European Coal Combustion Products Association, in 2004, the United States and Europe generated 64,500 and 43,500 kilotons of fly ash, respectively, of which only approximately 45 per cent was reused. One of the most common uses of coal fly ash is as an additive in making concrete while the remaining ash is often landfilled.

Although several studies have examined the chemical interactions of fly ash and contaminants, few studies have investigated the effects of fly ash on the bioavailability and toxicity of NOCs associated with sediments. In this study, the objective was to evaluate how effectively several samples of coal fly ash affect the toxicity and geochemistry of a marine sediment environmentally contaminated with PAHs. For this evaluation, a marine amphipod (*Ampelisca abdita*) and mysid (*Americamysis bahia*) were used.

This study is an evaluation of coal fly ash as a possible remedial substance for capping aquatic contaminated sites. For comparison, a form of activated carbon, powdered coconut charcoal, which has been shown to very effectively reduce the toxicity of NOCs in sediments, was included in the study. Finally, to assess the effects of fly ash addition on the nutritional condition of benthic organisms and uptake of possible contaminants, a preliminary 28-d bioaccumulation study with a marine polychaete (*Nereis virens*) was performed with the use of reference sediment amended with fly ash, and the effects on organism tissue and lipid mass and PAH and mercury accumulation were evaluated.

Selection of Fly Ash

Six fly ash samples collected from coal-fired power plants in the United States were evaluated. In general, for the experiments, fly ashes were used in a dry form. In limited cases, fly ashes were treated before use to evaluate effects on performance. Generally, these were fly ashes suspected of being effective at reducing toxicity because of their elevated loss-on-ignition values. One fly ash was treated by deaeration. Deaeration was performed to remove air inside the pores of the fly ash and improve the ability for contaminants to interact with the ash when wet. In the procedure, a mass of fly ash was combined with deionized water and mixed.

The mixture was placed in a flask under vacuum for approximately 18 h. The fly ash was then separated from the water by centrifugation. Powdered coconut charcoal PCB-G (FC) (90–96 per cent <45 ìm) was from Calgon Carbon (Pittsburgh, PA, USA) and also deaerated before use. In some other treatments, before use, a fly ash was rinsed with deionized water three times, the deionized water decanted, and replaced for storage with reconstituted seawater with a salinity of 30 (4°C in the dark). Reconstituted seawater was made by adding 100‰ brine prepared from Narragansett Bay seawater to deionized water.

Sediment and Fly Ash Additions

For toxicity testing, fly ash and coconut charcoal samples were combined with LIS and ER sediments at 20 per cent by weight, and for one fly ash also at 5 and 10 per cent (class F 65.5 per cent). For the preliminary bioaccumulation study, class F 65.5 per cent fly ash was added to LIS in 5 and 10 per cent by weight amendments.

Amendments were prepared by manually homogenizing sediments and fly ash or coconut charcoal with a stainless steel spatula. Amended sediments were then added to toxicity testing or bioaccumulation chambers. Because of the known high toxicity of ER sediment, a dilution effects treatment that used beach sand was not included.

Fly Ash Characteristics

Fly ashes varied in LOI with values ranging from 0.84 per cent for a class C fly ash to 65.5 per cent for a class F fly ash. Surface areas for the fly ashes followed the trend set with LOI: fly ashes with low LOI values had low surface areas (0.84 per cent and 3.46 m^2/g), and fly ashes with large LOI values had large surface areas (65.5 per cent and 35.3 m^2/g).

As expected, the foam index mirrored the LOI and surface area relationships; the fly ashes with the greatest LOIs also had elevated foam index values indicating poor fly ash for making concrete. In almost every case, the fly ashes had more micropore capacity than either mesopore or macropore volume. Furthermore, low-LOI fly ashes had much greater porosity than the high-LOI fly ashes, with values ranging from seven to 10 times greater depending on the pore size. Powdered coconut charcoal carbon content was approximately 73 per cent with a very large surface area of 1,100 m^2/g.

Finally, the class C fly ashes when added to the sediments were observed to cause them to solidify in their consistency. This quality of fly ash, called pozzolanicity (ability to react with calcium hydroxide to form solid-phase calcium silicate hydrates), is an advantage as a cement additive but is not necessarily desirable in an environmental remediation method.

Fly Ash Characterization

The six fly ash samples were evaluated for several geochemical characteristics, including loss on ignition (LOI), surface area, foam index, and porosity with the use of methods in Kulaots et al. Fly ashes formed from anthracite or bituminous coals were designated class F, whereas fly ashes from lignite or subbituminous coals were designated class C. For this report, fly ashes are identified by their class and LOI.

Unburned residual carbon content (LOI) was performed on 1-g samples of predried fly ash (130°C for 2 h) followed by 740°C for 2 h. Carbon content (percent C) of the coconut charcoal was determined on an elemental analyzer. Surface area was determined with nitrogen (N_2) in an automated Quantachrome Corporation Autosorb 1 gas adsorption instrument (Boynton Beach, FL, USA) according to the standard Brunauer, Emmet, and Teller (BET) analysis. Surface area for coconut charcoal was taken from Lebo et al.

Foam index is the commonly used field test for determining the suitability of a particular fly ash as a concrete additive. Generally, the larger the foam index, the poorer the fly ash is expected to perform as a concrete additive. Briefly, 2 g of fly ash, 8 g of Portland cement, and 25 ml of deionized water were combined in a 70-ml cylindrical jar (40 mm inner diameter), capped, and thoroughly shaken for 1 min to completely wet the cement and fly ash. Next, a 10 per cent aqueous solution of commercial surfactant air-entraining admixtures was titrated in 0.02-ml increments. After addition of each increment, the jar was capped and shaken for approximately 15 s, after which, the lid was removed and the liquid surface observed. The visual endpoint was the presence of a stable surface foam lasting at least 45 s. Fly ash porosity was categorized as micropores (<20 Å), mesopores (20–500 Å), and macropores (>500 Å) on the basis of the International Union of Pure and Applied Chemistry classification. To determine the porosity of each category, N_2 gas adsorption isotherms were measured at 77 K with the Autosorb-1 instrument described above.

Toxicity Testing

Toxicity test methods and acceptability criteria are discussed in detail by Ho et al. Briefly, testing involved adding 10 cultured 48-h-old mysids, *A. bahia,* and 10 field-collected (0.7–1.0 mm) amphipods, *A. abdita,* to each exposure chamber. Tests were performed without seawater renewal (static) with gentle aeration in 100-ml exposure chambers with 60 ml of reconstituted seawater added.

Reconstituted seawater was added gently to avoid resuspending the sediments. Organisms were added 24 h later. Mysids were fed daily 100 ml of a solution of 1.4 g (wet) newly hatched *Artemia* in 10 ml of deionized water, whereas the amphipods were not fed. After 7 d, the tests were terminated, and organisms were sieved from the exposure chambers and assessed for mortality (missing organisms were considered mortalities).

Preliminary Bioaccumulation Study

Exposures with LIS sediment with 5 and 10 per cent additions of class F 65.5 per cent fly ash were performed on the basis of guidance from the U.S. Environmental Protection Agency. Exposure chambers were 1-L beakers (15 × 12 cm) with a cylindrical coil of 3.0 mm mesh screening (12 × 10 × 0.1 cm) inserted in the top. Screening allowed for the flow-through seawater to overflow the beaker's edge while stopping organisms from swimming from the chamber.

Flow rates during the 28-d exposure were approximately 7 ml/min for approximately 10 volumetric turnovers per day. The flow-through exposure system was provided by a siphoning gravity diluter that distributed seawater to the chambers. Chambers contained 200 ml of sediment or sediment plus fly ash and one field-collected polychaete, *N. virens*. After addition to the test, animals that had not burrowed after 1 h were replaced. During the exposure, the organisms were not fed, and new sediment was not added.

For each treatment, three replicates each contained a worm with an initial weight of 2 to 5 g (wet wt). Sediments were added to chambers and the flow-through seawater system was started 3 d (day "3) before the animals were added. On day 0, preweighed animals were added to the exposures. Exposures were performed at 20°C and 30‰ under a 12:12 h light:dark cycle with 30-min brightening and dimming phases and continuous aeration.

On day 28, animals were removed from the exposure sediment by sieving through a 1-mm stainless steel sieve. Animals were then rinsed with 20°C seawater and placed into 400-ml static depuration chambers containing 40 g of LIS sediment and 150 ml of natural seawater. All chambers were aerated with one animal per chamber.

After approximately 24 h, the organisms were sieved out of the sediment, rinsed with seawater, weighed, and frozen at "4°C until chemical and lipid determinations. Lipids were determined gravimetrically from 50:50 acetone:hexane mixture extracts of the tissues. Three replicate extractions were performed with sonication on manually homogenized tissues. After the three extracts were composited, a subsample was collected and allowed to evaporate to dryness, and the lipid mass was determined.

Statistical Analyses

Treatments were performed with two to three replicates and the mean and standard deviation are generally reported. Toxicity testing, polychaete mass and lipid content, and bioaccumulation results were analyzed by analysis of variance followed by either a Dunnett's test or protected least significant difference test ($p = 0.05$). To normalize for the binomial distribution of percent survival data, the square root of the raw data was arcsine transformed before statistical analysis. For toxicity testing results, comparisons were performed as one-tailed tests to detect only treatments with significantly less survival than the reference (LIS) or greater survival than the toxic sediment (ER). Regression equations, as well as resulting coefficients of determination (r^2) and p values, were generated on untransformed toxicological and chemical data. Statistical analyses were performed with the SAS System, Release 8.02 and Statistica.

Chemical Analyses Methods

Chemical analyses were performed on overlying water samples collected on the last day of the toxicity testing exposures. Because of high flow rates in the bioaccumulation study, overlying water analyses were not conducted. Cationic metal analyses for cadmium, copper, nickel, and lead were performed on nitric acid–treated water samples by graphite furnace atomic spectroscopy. Detection limits for cationic metals were approximately 10, 25, 30, and 25 ìg/L for cadmium, copper, nickel, and lead, respectively.

In water and polychaete tissue samples, 13 PAHs were analyzed, including fluorene, phenanthrene, anthracene, fluoranthene, pyrene, benzanthracene, chrysene, benzopyrene, benzopyrene, perylene, indenopyrene, dibenzanthracene, and benzoperylene. In the water samples, PAHs were extracted three times with the use of acetone partitioned against deionized water and 1 ml of heptane.

For tissues, PAHs were extracted as described for the lipid analysis, except the extracts for PAH analysis were cleaned up in an AccuBond II solid phase extraction silica column and then reduced in volume to approximately 1 ml. Samples were stored at 4°C in the dark until analysis. Analyses were performed on a Hewlett-Packard 5890 Series II gas chromatograph with a 5971 mass spectrometer and a 60 m DB-5 fused silica capillary column.

Detection limits for PAHs were approximately 0.5 ng/L. Fly ash and whole-tissue samples were analyzed for mercury with a DMA-80 direct mercury analyzer. The instrument was calibrated with National Research Council of Canada standard reference materials for trace metals, including lobster heptopancreas (TORT-2) and dogfish muscle (DORM-2).

Carbon content of the powdered coconut charcoal was quantified on a ThermoFinnigan FlashEA Series 1112 automated elemental analyzer. Total ammonia (NH_x) was measured with an Orion ion-selective ammonia probe and meter. Detection limits for ammonia were approximately 0.1 mg/L. Along with ammonia, overlying water pH, salinity, and temperature were measured to calculate un-ionized ammonia (NH_3) according to Hampson. For chemical measures, quality assurance included performing analytical and sample blanks, duplicate or triplicate analyses, and, where appropriate, internal standard recoveries.

NUCLEAR WEAPONS REPROCESSING

Waste from nuclear weapons reprocessing (as opposed to production, which requires primary processing from reactor fuel) is unlikely to contain much beta or gamma activity other than tritium and americium. It is more likely to contain alpha emitting actinides such as Pu-239 which is a fissile material used in bombs, plus some material with much higher specific activities, such as Pu-238 or Po.

In the past the neutron trigger for a bomb tended to be beryllium and a high activity alpha emitter such as polonium; an alternative to polonium is Pu-238. For reasons of national security, details of the design of modern bombs are normally not released to the open literature. It is likely however that a D-T fusion reaction in either an electrically driven device or a D-T fusion reaction driven by the chemical explosives would be used to start up a modern device.

Some designs might well contain a radioisotope thermoelectric generator using Pu-238 to provide a longlasting source of electrical power for the electronics in the device.

It is likely that the fissile material of an old bomb which is due for refitting will contain decay products of the plutonium isotopes used in it, these are likely to include U-236 from Pu-240 impurities, plus some U-235 from decay of the Pu-239; however, due to the relatively long half-life of these Pu isotopes, these wastes from radioactive decay of bomb core material would be very small, and in any case, far less dangerous (even in terms of simple radioactivity) than the Pu-239 itself.

The beta decay of Pu-241 forms Am-241; the in-growth of americium is likely to be a greater problem than the decay of Pu-239 and Pu-240 as the americium is a gamma emitter (increasing external-exposure to workers) and is an alpha emitter which can cause the generation of heat. The plutonium could be separated from the americium by several different processes; these would include pyrochemical processes and aqueous/organic solvent extraction. A truncated PUREX type extraction process would be one possible method of making the separation.

Medical

Radioactive medical waste tends to contain beta particle and gamma ray emitters. It can be divided into two main classes. In diagnostic nuclear medicine a number of short-lived gamma emitters such as technetium-99m are used. Many of these can be disposed of by leaving it to decay for a short time before disposal as normal waste. Other isotopes used in medicine, with half-lives in parentheses:

- Y-90, used for treating lymphoma (2.7 days)
- I-131, used for thyroid function tests and for treating thyroid cancer (8.0 days)
- Sr-89, used for treating bone cancer, intravenous injection (52 days)
- Ir-192, used for brachytherapy (74 days)
- Co-60, used for brachytherapy and external radiotherapy (5.3 years)
- Cs-137, used for brachytherapy, external radiotherapy (30 years)

Industrial

Industrial source waste can contain alpha, beta, neutron or gamma emitters. Gamma emitters are used in radiography while neutron emitting sources are used in a range of applications, such as oil well logging.

Naturally occurring radioactive material (NORM)

Processing of substances containing natural radioactivity; this is often known as NORM. A lot of this waste is alpha particle-emitting matter from the decay chains of uranium and thorium. The main source of radiation in the human body is potassium-40 (^{40}K). There is a natural background radioactivity that life systems are built to resist. Most rocks, due to their components, have a certain, but low level, of radioactivity.

Coal

Coal contains a small amount of radioactive uranium, barium, thorium and potassium, but, in the case of pure coal, this is significantly less than the average concentration of those elements in the Earth's crust. However, the surrounding strata, if shale or mudstone, often contains slightly more than average and this may also be reflected in the ash content of 'dirty' coals. The more active ash minerals become concentrated in the fly ash precisely because they do not burn well.

However, the radioactivity of fly ash is still very low. It is about the same as black shale and is less than phosphate rocks, but is more of a concern because a small amount of the fly ash ends up in the atmosphere where it can be inhaled.

Oil and Gas

Residues from the oil and gas industry often contain radium and its daughters. The sulphate scale from an oil well can be very radium rich, while the water, oil and gas from a well often contains radon. The radon decays to form solid radioisotopes which form coatings on the inside of pipework. In an oil processing plant the area of the plant where propane is processed is often one of the more contaminated areas of the plant as radon has a similar boiling point as propane.

TYPES OF RADIOACTIVE WASTE

Although not significantly radioactive, uranium mill tailings are waste. They are byproduct material from the rough processing of uranium-bearing ore. They are sometimes referred to wastes, from the section of the U.S. Atomic Energy Act that defines them. Uranium mill tailings typically also contain chemically-hazardous heavy metals such as lead and arsenic. Vast mounds of uranium mill tailings are left at many old mining sites, especially in Colorado, New Mexico, and Utah.

Low level waste (LLW) is generated from hospitals and industry, as well as the nuclear fuel cycle. It comprises paper, rags, tools, clothing, filters, etc., which contain small amounts of mostly short-lived radioactivity. Commonly, LLW is designated as such as a precautionary measure if it originated from any region of an 'Active Area', which frequently includes offices with only a remote possibility of being contaminated with radioactive materials. Such LLW typically exhibits no higher radioactivity than one would expect from the same material disposed of in a non-active area, such as a normal office block. Some high activity LLW requires shielding during handling and transport but most LLW is suitable for shallow land burial. To reduce its volume, it is often compacted or incinerated before disposal. Low level waste is divided into four classes, class A, B, C and GTCC, which means "Greater Than Class C".

Intermediate level waste (ILW) contains higher amounts of radioactivity and in some cases requires shielding. ILW includes resins, chemical sludge and metal reactor fuel cladding, as well as contaminated materials from reactor decommissioning. It may be solidified in concrete or bitumen for disposal. As a general rule, short-lived waste (mainly non-fuel materials from reactors) is buried in shallow repositories, while long-lived waste (from fuel and fuel-reprocessing) is deposited in deep underground facilities. U.S. regulations do not define this category of waste; the term is used in Europe and elsewhere.

High level waste (HLW) is produced by nuclear reactors. It contains fission products and transuranic elements generated in the reactor core. It is highly radioactive and often thermally hot. LLW and ILW accounts for over 95% of the total radioactivity produced in the process of nuclear electricity generation. The amount of HLW worldwide is currently increasing by about 12,000 metric tons every year, which is the equivalent to about 100 double-decker busses or a two-story structure built on top of a basketball court.

Transuranic waste (TRUW) as defined by U.S. regulations is, without regard to form or origin, waste that is contaminated with alpha-emitting transuranic radionuclides with half-lives greater than 20 years, and concentrations greater than 100 nCi/g (3.7 MBq/kg), excluding High Level Waste. Elements that have an atomic number greater than uranium are called transuranic ("beyond uranium").

Because of their long half-lives, TRUW is disposed more cautiously than either low level or intermediate level waste. In the U.S. it arises mainly from weapons production, and consists of clothing, tools, rags, residues, debris and other items contaminated with small amounts of radioactive elements (mainly plutonium).

Under U.S. law, TRUW is further categorized into "contact-handled" (CH) and "remote-handled" (RH) on the basis of radiation dose measured at the surface of the waste container. CH TRUW has a surface dose rate not greater than 200 mrem per hour (2 mSv/h), whereas RH TRUW has a surface dose rate of 200 mrem per hour (2 mSv/h) or greater. CH TRUW does not have the very high radioactivity of high level waste, nor its high heat generation, but RH TRUW can be highly radioactive, with surface dose rates up to 1000000 mrem per hour (10000 mSv/h).

The United States currently permanently disposes of TRUW generated from nuclear power plants and military facilities at the Waste Isolation Pilot Plant.

Management of waste

Of particular concern in nuclear waste management are two long-lived fission products, Tc-99 (half-life 220,000 years) and I-129 (half-life 17 million years), which dominate spent fuel radioactivity after a few thousand years. The most troublesome transuranic elements in spent fuel are Np-237 (half-life two million years) and Pu-239 (half life 24,000 years). Nuclear waste requires sophisticated treatment and management in order to successfully isolate it from interacting with the biosphere.

This usually necessitates treatment, followed by a long-term management strategy involving storage, disposal or transformation of the waste into a non-toxic form. Governments around the world are considering a range of waste management and disposal options, though there has been limited progress toward long-term waste management solutions.

Initial treatment of waste

Long-term storage of radioactive waste requires the stabilization of the waste into a form which will not react, nor degrade, for extended periods of time. One way to do this is through vitrification. Currently at Sellafield the high-level waste (PUREX first cycle raffinate) is mixed with sugar and then calcined. Calcination involves passing the waste through a heated, rotating tube. The purposes of calcination are to evaporate the water from the waste, and de-nitrate the fission products to assist the stability of the glass produced.

The 'calcine' generated is fed continuously into an induction heated furnace with fragmented glass. The resulting glass is a new substance in which the waste products are bonded into the glass matrix when it solidifies. This product, as a molten fluid, is poured into stainless steel cylindrical containers ("cylinders") in a batch process. When cooled, the fluid solidifies ("vitrifies") into the glass. Such glass, after being formed, is very highly resistant to water. After filling a cylinder, a seal is welded onto the cylinder. The cylinder is then washed. After being inspected for external contamination, the steel cylinder is stored, usually in an underground repository. In this form, the waste products are expected to be immobilized for a very long period of time (many thousands of years).

The glass inside a cylinder is usually a black glossy substance. All this work (in the United Kingdom) is done using hot cell systems. The sugar is added to control the ruthenium chemistry and to stop the formation of the volatile RuO_4 containing radio ruthenium. In the west, the glass is normally a borosilicate glass (similar to Pyrex), while in the former Soviet bloc it is normal to use a phosphate glass. The amount of fission products in the glass must be limited because some (palladium, the other Pt group metals, and tellurium) tend to form metallic phases which separate from the glass. In Germany a vitrification plant is in use; this is treating the waste from a small demonstration reprocessing plant which has since been closed down.

Synroc

The Australian Synroc (synthetic rock) is a more sophisticated way to immobilize such waste, and this process may eventually come into commercial use for civil wastes (it is currently being developed for U.S. military wastes). Synroc was invented by the late Prof Ted Ringwood (a geochemist) at the Australian National University. The Synroc contains pyrochlore and cryptomelane type minerals. The original form of Synroc (Synroc C) was designed for the liquid high level waste (PUREX raffinate) from a light water reactor. The main minerals in this Synroc are hollandite ($BaAl_2Ti_6O_{16}$), zirconolite ($CaZrTi_2O_7$) and perovskite ($CaTiO_3$). The zirconolite and perovskite are hosts for the actinides. The strontium and barium will be fixed in the perovskite. The caesium will be fixed in the hollandite.

Ion exchange

It is common for medium active wastes in the nuclear industry to be treated with ion exchange or other means to concentrate the radioactivity into a small volume. The much less radioactive bulk (after treatment) is often then discharged. For instance, it is possible to use a ferric hydroxide floc to remove radioactive metals from aqueous mixtures. After the radioisotopes are absorbed onto the ferric hydroxide, the resulting sludge can be placed in a metal drum before being mixed with cement to form a solid waste form. In order to get better long-term performance (mechanical stability) from such forms, they may be made from a mixture of fly ash, or blast furnace slag, and portland cement, instead of normal concrete (made with portland cement, gravel and sand).

Long term management of waste

The timeframe in question when dealing with radioactive waste ranges from 10,000 to 1,000,000 years, according to studies based on the effect of estimated radiation doses.Researchers suggest that forecasts of health detriment for such periods *should be examined critically*. Practical studies only consider up to 100 years as far as effective planning and cost evaluations are concerned. Long term behaviour of radioactive wastes remains a subject for ongoing research projects.

Geologic disposal

The process of selecting appropriate deep final repositories for high level waste and spent fuel is now under way in several countries with the first expected to be commissioned some time after 2010. The basic concept is to locate a large, stable geologic formation and use mining technology to excavate a tunnel, or large-bore tunnel boring machines (similar to those used to drill the Chunnel from England to France) to drill a shaft 500–1,000 meters below the surface where rooms or vaults can be excavated for disposal of high-level radioactive waste. The goal is to permanently isolate nuclear waste from the human environment. However, many people remain uncomfortable with the immediate stewardship cessation of this disposal system, suggesting perpetual management and monitoring would be more prudent.

Because some radioactive species have half-lives longer than one million years, even very low container leakage and radionuclide migration rates must be taken into account. Moreover, it may require more than one half-life until some nuclear materials lose enough radioactivity to no longer be lethal to living things. A 1983 review of the Swedish radioactive waste disposal program by the National Academy of Sciences found that country's estimate of several hundred thousand years—perhaps up to one million years—being necessary for waste isolation "fully justified."

Storing high level nuclear waste above ground for a century or so is considered appropriate by many scientists. This allows the material to be more easily observed and any problems detected and managed, while decay of radionuclides over this time period significantly reduces the level of radioactivity and associated harmful effects to the container material. It is also considered likely that over the next century newer materials will be developed which will not break down as quickly when exposed to a high neutron flux, thus increasing the longevity of the container once it is permanently buried.

Sea-based options for disposal of radioactive waste include burial beneath a stable abyssal plain, burial in a subduction zone that would slowly carry the waste downward into the Earth's mantle, and burial beneath a remote natural or human-made island.

While these approaches all have merit and would facilitate an international solution to the vexing problem of disposal of radioactive waste, they are currently not being seriously considered because of the legal barrier of the Law of the Sea and because in North America and Europe sea-based burial has become taboo from fear that such a repository could leak and cause widespread damage. Dumping of radioactive waste from ships has reinforced this concern, as has contamination of islands in the Pacific Ocean. However, sea-based approaches might come under consideration in the future by individual countries or groups of countries that cannot find other acceptable solutions.

The proposed land-based subductive waste disposal method disposes of nuclear waste in a subduction zone accessed from land, and therefore is not prohibited by international agreement. This method has been described as the most viable means of disposing of radioactive waste, and as the state-of-the-art in nuclear waste disposal technology. Another approach termed Remix & Return would blend high-level waste with uranium mine and mill tailings down to the level of the original radioactivity of the uranium ore, then replace it in inactive uranium mines. This approach has the merits of providing jobs for miners who would double as disposal staff, and of facilitating a cradle-to-grave cycle for radioactive materials. However, this approach would be inappropriate for spent reactor fuel in the absence of reprocessing, due to the presence in it of highly toxic radioactive elements such as plutonium.

Re-use of waste

Another option is to find applications of the isotopes in nuclear waste so as to re-use them. Already, caesium-137, strontium-90 and a few other isotopes are extracted for certain industrial applications such as food irradiation and radioisotope thermoelectric generators. While re-use does not eliminate the need to manage radioisotopes, it may reduce the quantity of waste produced.

The Nuclear Assisted Hydrocarbon Production Method, Canadian patent application 2,638,179, is a method for the temporary or permanent storage of nuclear waste materials comprising the placing of waste materials into one or more repositories or boreholes constructed into an unconventional oil formation. The thermal flux of the waste materials fracture the formation, alters the chemical and/or physical properties of hydrocarbon material within the subterranean formation to allow removal of the altered material. A mixture of hydrocarbons, hydrogen, and/or other formation fluids are produced from the formation. The radioactivity of high-level radioactive waste affords proliferation resistance to plutonium placed in the periphery of the repository or the deepest portion of a borehole.

Space disposal

Space disposal is an attractive notion because it permanently removes nuclear waste from the environment. However, it has significant disadvantages, not least of which is the potential for catastrophic failure of a launch vehicle. Furthermore, the high number of launches that would be required — due to the fact that no individual rocket would be able to carry very much of the material relative to the material needed to be disposed of—makes the proposal impractical (for both economic and risk-based reasons). To further complicate matters, international agreements on the regulation of such a program would need to be established. This method would also be energetically intensive and thus is not necessarily economically feasible.

It has been suggested that through the use of a stationary launch system many of the risks of catastrophic launch failure could be avoided. A promising concept is the use of high power lasers to launch "indestructible" containers from the ground into space. Such a system would require no rocket propellant, with the launch vehicle's payload making up a near entirety of the vehicle's mass. Without the use of rocket fuel on board there would be little chance of the vehicle exploding. Another form of safe removal would possibly be the space elevator. Encasing the waste in glassified form inside a steel shell 9 inches (230 mm) thick, which in turn is tiled with shuttle tile to its exterior. If the launch vehicle fails just before reaching orbit, the waste ball will safely re-enter the Earth's atmosphere. The steel shell would deform on impact, but would not rupture due to the density of the shell. Also, this would potentially allow the waste to be shot into the Sun.

Transmutation

There have been proposals for reactors that consume nuclear waste and transmute it to other, less-harmful nuclear waste. In particular, the Integral Fast Reactor was a proposed nuclear reactor with a nuclear fuel cycle that produced no transuranic waste and in fact, could consume transuranic waste. It proceeded as far as large-scale tests but was then canceled by the U.S. Government. Another approach, considered safer but requiring more development, is to dedicate subcritical reactors to the transmutation of the left-over transuranic elements. Transmutation was banned in the United States on April 1977 by President Carter due to the danger of plutonium proliferation, but President Reagan rescinded the ban in 1981. Due to the economic losses and risks, construction of reprocessing plants during this time did not resume. Due to high energy demand, work on the method has continued in the EU. This has resulted in a practical nuclear research reactor called Myrrha in which transmutation is possible. Additionally, a new research program called ACTINET has been started in the EU to make transmutation possible on a large, industrial scale.

There have also been theoretical studies involving the use of fusion reactors as so called "actinide burners" where a fusion reactor plasma such as in a tokamak, could be "doped" with a small amount of the "minor" transuranic atoms which would be transmuted (meaning fissioned in the actinide case) to lighter elements upon their successive bombardment by the very high energy neutrons produced by the fusion of deuterium and tritium in the reactor. It was recently found by a study done at MIT, that only 2 or 3 fusion reactors with parameters similar to that of the International Thermonuclear Experimental Reactor (ITER) could transmute the entire annual minor actinide production from all of the light water reactors presently operating in the United States fleet while simultaneously generating approximately 1 gigawatt of power from each reactor.

National management plans

Most countries are considerably behind the United States in developing plans for high-level radioactive waste disposal. Sweden and Finland are furthest along in committing to a particular disposal technology, while many others reprocess spent fuel or contract with France or Great Britain to do it, taking back the resulting plutonium and high-level waste. "An increasing backlog of plutonium from reprocessing is developing in many countries. It is doubtful that reprocessing makes economic sense in the present environment of cheap uranium."

In many European countries (e.g., Britain, Finland, the Netherlands, Sweden and Switzerland) the risk or dose limit for a member of the public exposed to radiation from a future high-level nuclear waste facility is considerably more stringent than that suggested by the International Commission on Radiation Protection or proposed in the United States. European limits are often more stringent than the standard suggested in 1990 by the International Commission on Radiation Protection by a factor of 20, and more stringent by a factor of ten than the standard proposed by the U.S. Environmental Protection Agency (EPA) for Yucca Mountain for the first 10,000 years after closure. Moreover, the U.S. EPA's proposed standard for greater than 10,000 years is 250 times more permissive than the European limit.

Accidents involving radioactive waste

A number of incidents have occurred when radioactive material was disposed of improperly, shielding during transport was defective, or when it was simply abandoned or even stolen from a waste store. In the Soviet Union, waste stored in Lake Karachay was blown over the area during a dust storm after the lake had partly dried out. At Maxey Flat, a low-level radioactive waste facility located in Kentucky, containment trenches covered with dirt, instead of steel or cement, collapsed under heavy rainfall into the trenches and filled with water.

The water that invaded the trenches became radioactive and had to be disposed of at the Maxey Flat facility itself. In other cases of radioactive waste accidents, lakes or ponds with radioactive waste accidentally overflowed into the rivers during exceptional storms.

Scavenging of abandoned radioactive material has been the cause of several other cases of radiation exposure, mostly in developing nations, which may have less regulation of dangerous substances (and sometimes less general education about radioactivity and its hazards) and a market for scavenged goods and scrap metal.

The scavengers and those who buy the material are almost always unaware that the material is radioactive and it is selected for its aesthetics or scrap value. Irresponsibility on the part of the radioactive material's owners, usually a hospital, university or military, and the absence of regulation concerning radioactive waste, or a lack of enforcement of such regulations, have been significant factors in radiation exposures.

Transportation accidents involving spent nuclear fuel from power plants are unlikely to have serious consequences due to the strength of the spent nuclear fuel shipping casks.

In Italy, several radioactive waste deposits let material flow into river water, thus contaminating water fit for domestic use.

EFFECTS OF FLY ASH ON SEDIMENT TOXICITY

Mysid survival in the LIS reference sediments without any fly ash added ranged from 87 to 97 per cent. Conversely, in some cases, mysid survival in LIS sediment amended with fly ashes showed significant toxicity. For example, exposure to class C 0.84 per cent and 1.3 per cent, as well as the class F 33.6 per cent and class F 35 per cent fly ashes, resulted in statistically significant reductions in survival in LIS. Both of the class C fly ashes were toxic, suggesting that some characteristic of this type of ash, possibly related to pozzolanicity, causes toxicity.

Both of these organisms interact with sediment surfaces, anything that reduces that interaction, like the sediment hardening caused by fly ash pozzolanicity, could result in additional stress. The class F 6.1 per cent, class F 33.6 per cent deaerated, and all doses of the class F 65.5 per cent fly ash (5, 10, and 20 per cent) showed no significant toxicity. The coconut charcoal treatment demonstrated nearly 100 per cent survival. Exposure of the mysid to ER and ER with fly ash amendments resulted in significant toxicity in many treatments. However, the presence of fly ash with the ER sediment reduced toxicity in all doses of the class F 35 per cent and 65.5 per cent fly ashes and coconut charcoal treatments. Toxicity reductions in the rinsed class F 65.5 per cent treatments were only marginal. In contrast, in the 5 and 10 per cent (nonrinsed) and 20 per cent class F 65.5 per cent, 20 per cent class F 35 per cent, and coconut charcoal amendments, toxicity reductions were significant compared with the ER nonamended treatments. In all other treatments, the fly ash did not significantly reduce toxicity to the mysid, including the rinsed treatments.

Amphipod survival in the LIS sediment was between 97 and 100 per cent. Adverse effects of the fly ashes on amphipod survival in LIS sediment were less severe than on the mysid. When compared with the LIS treatments, the only statistically significantly toxic fly ashes were the class F 33.6 per cent and class F 65.5 per cent (20 per cent amendment only). However, the deaerated form of class F 33.6 per cent was not toxic (as also observed with the mysid). As with the mysid, the coconut charcoal did not cause any significant toxicity to the amphipod. In contrast, exposure of the amphipod to ER and ER with fly ash amendments caused significant toxicity in several treatments. Presence of three fly ashes, including class F 33.6 per cent, class F 35 per cent, and all doses of class F 65.5 per cent (except the 5 per cent rinsed treatment), resulted in a significant decrease in toxicity compared with the nonamended ER sediments. The coconut charcoal also removed a significant portion of observed toxicity caused by ER sediment.

Despite the observed mortality associated with some of the fly ashes, compared with LIS, when the survival data for both species in the ER treatments are presented versus the amount of carbon added to each exposure chamber (with LOI or elemental analysis data), a linear, significant ($r^2 = 0.56$) relationship is found. These data, along with the research of others discussed earlier, suggest the carbon associated with the fly ash is acting as a form of highly adsorbent black carbon that sequesters the PAHs to the sediment–fly ash mixture and reduces bioavailability as manifested here by toxicity.

Consequently, considerations for using fly ash as a contaminated sediment remediation amendment should involve ashes with the higher carbon contents. However, the high-carbon fly ashes were often toxic when tested with our reference sediment. Understanding the potential causes of this toxicity will be needed to use fly ash in a viable remedial application.

Finally, the amendment of fly ash to the LIS and ER sediments did result in an increase in overlying water pH in the toxicity testing chambers. In the LIS sediments, the increase in pH averaged 0.17, 0.59, and 0.93 for the 5, 10, and 20 per cent additions of the class F 65.5 per cent fly ash, respectively. In the ER sediment, the increases were smaller, with changes of 0.24, 0.42, and 0.34 for the 5, 10, and 20 per cent additions of class F 65.5 per cent fly ash, respectively.

For the other fly ashes added at 20 per cent, the mean pH increase in the overlying waters of the LIS treatments was 1.17 ± 0.15, and the increase in the ER treatments was 0.16 ± 0.22. Although these changes in pH are unlikely to be toxic by themselves to the mysid and amphipod used in this study, the increases in pH might have implications for the toxicity of any pH-dependent toxicants, like ammonia, associated with the fly ashes. This could contribute to understanding the causes of apparent fly ash toxicity observed in the LIS reference sediment amendments.

Effects of fly Ash on Overlying Water PAHs

Addition of fly ash nearly universally reduced the concentrations of PAHs measured in the overlying water of the exposure chambers after 7 d. In the LIS reference sediment treatments (data not shown), the sediment alone resulted in very low levels of five PAHs being detected in the overlying waters, with all at concentrations less than 4 ìg/L. In the class C 1.3 per cent treatment, three PAHs were measured in the overlying water, but at concentrations of 1 ìg/L or less.

In the other treatments, overlying water PAH concentrations were below detection. Given the relatively low concentrations of PAHs typically found in LIS sediment, these overlying water concentrations were expected.

Overall, in the ER treatments with fly ash present, as the PAHs increased in molecular weight, the amount present in the overlying water decreased compared with the ER treatment without fly ash (or coconut charcoal). More specifically, in the ER treatments, the observed trend was that as LOI values increased, the concentration of PAHs in the overlying water decreased. Consequently, in the class F 35 per cent and 65.5 per cent fly ashes, few PAHs were detected in the overlying water, and at levels less than 1 ìg/L.

Similarly, concentrations of PAHs in the overlying water of class F 33.6 per cent fly ash (regular and deaerated) were relatively low compared with the low-LOI fly ashes. Provides the overlying water concentration of total PAHs (sum of the 13 PAHs measured) versus the amount of carbon added to the exposure chambers. The concentration of total PAHs is shown to decrease as the amount of carbon increases, resulting in a strong correlation ($r^2 = 0.88$) and significant relationship. The class C fly ashes demonstrated, in some cases, elevated concentrations of low– and medium–molecular weight PAHs in the overlying water, especially fluoranthene and pyrene, which could indicate that class C fly ash is releasing PAHs.

However, previous analysis of class C fly ashes in our laboratory found very little or no PAHs to be extractable. This type of possible release was not observed with any of the class F fly ashes or in the LIS treatments. It might also be possible that the ER water extracts from treatments amended with class C fly ashes were contaminated with small sediment particles, artificially elevating the overlying water concentrations. However, the concentrations of the higher molecular weight PAHs in these samples do not follow the trend expected if this were the case when compared with the treatment containing only ER sediments.

Unlike the PAHs, the cationic metals cadmium, copper, nickel, and lead were not detected in the overlying water of any sediment treatments. Given the relatively low concentrations of cationic metals associated with both LIS and ER, the lack of elevated concentrations in the overlying water was expected.

Furthermore, leaching of cationic metals into the overlying water from the fly ash–treated samples was not observed, suggesting, at least preliminarily, that the fly ashes did not represent a risk of toxic exposure for these metals. Other metals associated with fly ash include mercury, arsenic, and selenium.

When evaluating the possible adverse effects of fly ash, further research on the release of these metals is recommended.

Suspected Effects of Ammonia on Toxicity

The presence of ammonia might explain the toxicity observed with some of the fly ashes when added to the LIS reference sediment. Two fly ashes were from coal-fired power plants that use selective catalytic reduction (SCR) or selective noncatalytic reduction (SNCR) technologies to control acid rain–causing nitric oxide emissions.

These technologies can also result in ammonia residues, including ammonium sulfates, associated with the organic and inorganic components of the fly ash. Class F fly ashes with 33.6 per cent and 65.5 per cent LOI were treated with these technologies. In general, the amount of total ammonia measured in the overlying waters of the exposure chambers on the last day of the test increased with increasing LOI values.

For example, in the LIS treatments, the class F fly ashes with 33.6 per cent and 65.5 per cent LOI have total ammonia concentrations of 12.8 and 15 mg/L, respectively. Both of these were treated with SCR or SNCR and are among the more toxic fly ashes when added to the LIS sediment. Although the mysids found the class C fly ashes just as toxic.

In contrast, the class F 35 per cent fly ash was not treated with SCR or SNCR and had relatively low toxicity and total ammonia concentrations in the LIS treatment despite having high LOI. This fly ash was also effective at reducing the toxicity of the ER sediment. Interestingly, the deaeration process resulted in the class F 33.6 per cent sediment having greatly reduced overlying water ammonia concentrations and being less toxic to both species.

In aqueous solution, the toxic component of total ammonia is ammonia (NH_3), a nonionic substance toxic to both the mysid and amphipod, with the median lethal concentration ranging from approximately 0.5 to 5.0 mg/L. Long Island Sound reference sediment overlying water NH_3 concentrations for fly ash treatments with 33.6 per cent and 65.5 per cent LOI easily extended into this range with values from 0.9 to 1.1 mg/L. Survival versus overlying water NH_3 concentrations, with a strong and significant relationship between amphipod survival and NH_3 ($r^2 = 0.91$).

The relationship for mysids was less strong ($r^2 = 0.23$) and insignificant ($p = 0.07$). This apparent disagreement between the species' responses could reflect the amphipod's exposure via the sediments, where the fly ash ammonia is present at the greatest concentrations, and the mysid's exposure via the water column, where the NH_3 concentrations can be, by comparison, diluted. Furthermore, the mysids could be responding to an unidentified toxicant.

The deaeration process used in this study involved placing the fly ash, or coconut charcoal, in a deionized water solution under a vacuum for approximately 18 h. It would appear that the vacuum treatment effectively removes toxic NH_3 from the fly ash. Gao et al. evaluated several other dry and semidry methods for removing ammonia from fly ash, including static and flowing humid air, water aerosols (fog), and ozonation with and without chemical oxidation. Methods that used flowing humid air and water aerosols showed the most promise in reducing ammonia concentrations. In this study, use of water to simply rinse the fly ash also appeared to reduce ammonia concentrations.

Unfortunately, both the deaeration and rinsing processes appeared to limit the effectiveness of the fly ashes at reducing the toxicity of the contaminated sediment. Toxicity data from the current study indicates the use of fly ashes, especially SCR and SNCR treated, as capping material at contaminated sites would require the ash be cleaned of ammonia in some fashion not affecting its ability to reduce contaminant bioavailability before application. However, higher LOI fly ashes not SCR or SNCR treated, like the class F 35 per cent used in this study, are also available.

Finally, despite the suggested relationship between observed toxicity and the presence of ammonia in some fly ash treatments, other toxicants could be active. Specifically, these other toxicants could include water soluble organic contaminants, metals, or both. Furthermore, the vacuum deaeration process used to wet the fly ash along with the deionized water rinsing would, in principle, also remove potentially toxic volatile organic contaminants. The roles of volatile organic contaminants and other toxicants in causing fly ash–related toxicity need to be examined further.

Effects of Fly Ash on Polychaete Mass

Some concern has been expressed about the implications of adding carbon-rich materials to environmental sediments because of possible adverse effects to indigenous organisms. For example, Jonker et al. reported reductions in the lipid mass of oligochaete worms (*Limnodrilus* sp.) exposed to PCB-contaminated sediments amended with anthracite coal and charcoal. Furthermore, research by Millward et al.

Reductions in PCB bioaccumulation by the polychaete *Neanthes arenaceodentata*, when activated carbon was amended into contaminated field sediments, the worm demonstrated reduced growth rates. Finally, McLeod et al. Reported that bivalves (*Macoma balthica*) showed lower condition indices when exposed to PCB-contaminated field sediments amended with activated carbon. These findings suggest that the addition of certain carbon-rich materials could have adverse ecological effects.

In this study, addition of fly ash at 5 and 10 per cent doses did not significantly affect worm growth during the 28-d exposures. The reference sediment showed a larger change in growth, but not significantly. Interestingly, the presence of fly ash resulted in an increasing trend in lipid content in the polychaetes as fly ash dose increased, but the trend was not significant despite the tripling of lipid mass in the 10 per cent fly ash treatment compared with the reference sediments. These preliminary data suggest that the presence of coal fly ash does not significantly reduce growth or lipid content in worms. Because the bioaccumulation data are preliminary, further evaluation of the effects of coal fly ash on organism growth and lipid content are necessary to definitively address this question.

Fly ash can contain several potentially toxic contaminants as residues resulting from the burning of coal. These contaminants can include PAHs and mercury. The presence or possible presence of these contaminants has generated concerns about use of fly ashes for environmental remediation. Analysis of worm tissues for PAHs in this study found none were detectable. This finding is not unexpected for at least two reasons. First, several studies have shown that polychaetes, like *N. virens*, have the ability to metabolize PAHs efficiently.

In future research, to get a better assessment of the bioavailability of PAHs from a fly ash exposure, use of a non-PAH–metabolizing species is recommended. Second, PAHs, when associated with fly ash, are extremely difficult to remove, even with the use of laboratory solvent extraction techniques. In addition, Vaporil et al. reported the results of the use of gut juice of the estuarine lugworm *Arenicola marina* to extract PAHs from several forms of black carbon, including diesel soot, coal dust, and a class F fly ash. Gut juice was unable to extract any PAHs from the fly ash, whereas several of the other forms of black carbon did release PAHs.

For the fly ashes evaluated in this study, concentrations of mercury ranged from approximately 4 to 1,500 ìg/kg. However, analysis of mercury in worm tissues resulted in a range of 175 to 271 ìg/kg (dry), with the LIS sediments having the greatest concentrations followed by the 10 per cent fly ash addition. Although published mercury bioaccumulation data for polychaetes is limited, these values are similar to those reported in the polychaete *Maldanes sarsi* collected from relatively pristine waters of the Barents Sea: 130 to 450 ìg/kg (dry).

Overall, despite the elevated levels of mercury in the class F 65.5 per cent fly ash, the data suggested mercury was not readily released during the 28-d bioaccumulation study. These data, although preliminary, agree with leaching studies in which mercury was not shown to readily release from fly ash into the aqueous phase even when very high concentrations of ammonia (1,000 mg NH_3/L) were used to enhance mercury dissolution from the ash.

RADIOACTIVE WASTE MANAGEMENT

All parts of the nuclear fuel cycle produce some radioactive waste (radwaste) and the relatively modest cost of managing and disposing of this is part of the electricity cost, i.e. it is internalised and paid for by the electricity consumers.

At each stage of the fuel cycle there are proven technologies to dispose of the radioactive wastes safely. For low- and intermediate-level wastes these are mostly being implemented. For high-level wastes some countries await the accumulation of enough of it to warrant building geological repositories; others, such as the USA, have encountered political delays.

Unlike other industrial wastes, the level of hazard of all nuclear waste - its radioactivity - diminishes with time. Each radionuclidea contained in the waste has a half-life – the time taken for half of its atoms to decay and thus for it to lose half of its radioactivity. Radionuclides with long half-lives tend to be alpha and beta emitters – making their handling easier – while those with short half-lives tend to emit the more penetrating gamma rays. Eventually all radioactive wastes decay into non-radioactive elements. The more radioactive an isotope is, the faster it decays.

The main objective in managing and disposing of radioactive (or other) waste is to protect people and the environment. This means isolating or diluting the waste so that the rate or concentration of any radionuclides returned to the biosphere is harmless. To achieve this, practically all wastes are contained and managed – some clearly need deep and permanent burial. From nuclear power generation, none is allowed to cause harmful pollution.

Radioactive wastes are waste types containing radioactive chemical elements that do not have a practical purpose. They are usually the products of nuclear processes, such as nuclear fission. However, industries not directly connected to the nuclear industry may produce large quantities of radioactive waste. The majority of radioactive waste is "low-level waste", meaning it contains low levels of radioactivity per mass or volume. This type of waste often consists of used protective clothing, which is only slightly contaminated but still dangerous in case of radioactive contamination of a human body through ingestion, inhalation, absorption, or injection.

The issue of disposal methods for nuclear waste was one of the most pressing current problems the international nuclear industry faced when trying to establish a long term energy production plan, yet there was hope it could be safely solved.

A recent research report on the Nuclear Industry perspective of the current state of scientific knowledge in predicting the extent that waste would find its way from the deep burial facility - back to soil and drinking water (such that it presents a direct threat to the health of human beings - as well as to other forms of life) is presented in a document from the IAEA (The International Atomic Energy Agency).

This document states "The capacity to model all the effects involved in the dissolution of the waste form, in conditions similar to the disposal site, is the final goal of all the research undertaken by many research groups over many years."

In the United States, DOE acknowledges progress in addressing the waste problems of the industry, and successful remediation of some contaminated sites, yet also major uncertainties and sometimes complications and setbacks in handling the issue properly, cost effectively, and in the projected time frame. In other countries with lower ability or will to maintain environmental integrity the issue would be even more problematic.

Nature of Radioactive Waste

Radioactive waste typically comprises a number of radioisotopes: unstable configurations of elements that decay, emitting ionizing radiation which can be harmful to human health and to the environment. Those isotopes emit different types and levels of radiation, which last for different periods of time.

The radioactivity of all nuclear waste diminishes with time. All radioisotopes contained in the waste have a half-life - the time it takes for any radionuclide to lose half of its radioactivity and eventually all radioactive waste decays into non-radioactive elements.

Certain radioactive elements (such as plutonium-239) in "spent" fuel will remain hazardous to humans and other living beings for hundreds of thousands of years. Other radioisotopes will remain hazardous for millions of years. Thus, these wastes must be shielded for centuries and isolated from the living environment for hundreds of millennia. Some elements, such as Iodine-131, have a short half-life (around 8 days in this case) and thus they will cease to be a problem much more quickly than other, longer-lived, decay products but their activity is much greater initially. The two tables show some of the major radioisotopes, their half-lives, and their radiation yield as a proportion of the yield of fission of Uranium-235.

The faster a radioisotope decays, the more radioactive it will be. The energy and the type of the ionizing radiation emitted by a pure radioactive substance are important factors in deciding how dangerous it will be. The chemical properties of the radioactive element will determine how mobile the substance is and how likely it is to spread into the environment and contaminate human bodies. This is further complicated by the fact that many radioisotopes do not decay immediately to a stable state but rather to a radioactive decay product leading to decay chains.

Exposure to high levels of radioactive waste may cause serious harm or death. Treatment of an adult animal with radiation or some other mutation-causing effect, such as a cytotoxic anti-cancer drug, may cause cancer in the animal. In humans it has been calculated that a 5 sievert dose is usually fatal, and the lifetime risk of dying from radiation induced cancer from a single dose of 0.1 sieverts is 0.8%, increasing by the same amount for each additional 0.1 sievert increment of dosage. Ionizing radiation causes deletions in chromosomes. If a developing organism such as an unborn child is irradiated, it is possible a birth defect may be induced, but it is unlikely this defect will be in a gamete or a gamete forming cell. The incidence of radiation-induced mutations in humans is undetermined, due to flaws in studies done to date. Depending on the decay mode and the pharmacokinetics of an element (how the body processes it and how quickly), the threat due to exposure to a given activity of a radioisotope will differ. For instance Iodine-131 is a short-lived beta and gamma emitter but because it concentrates in the thyroid gland, it is more able to cause injury than cesium-137 which, being water soluble, is rapidly excreted in urine. In a similar way, the alpha emitting actinides and radium are considered very harmful as they tend to have long biological half-lives and their radiation has a high linear energy transfer value. Because of such differences, the rules determining biological injury differ widely according to the radioisotope, and sometimes also the nature of the chemical compound which contains the radioisotope.

Nuclear fuel cycle

Front End

Waste from the front end of the nuclear fuel cycle is usually alpha emitting waste from the extraction of uranium. It often contains radium and its decay products.

Uranium dioxide (UO_2) concentrate from mining is not very radioactive - only a thousand or so times as radioactive as the granite used in buildings. It is refined from yellowcake (U_3O_8), then converted to uranium hexafluoride gas (UF_6). As a gas, it undergoes enrichment to increase the U-235 content from 0.7% to about 4.4% (LEU). It is then turned into a hard ceramic oxide (UO_2) for assembly as reactor fuel elements.

The main by-product of enrichment is depleted uranium (DU), principally the U-238 isotope, with a U-235 content of ~0.3%. It is stored, either as UF_6 or as U_3O_8. Some is used in applications where its extremely high density makes it valuable, such as the keels of yachts, and anti-tank shells. It is also used (with recycled plutonium) for making mixed oxide fuel (MOX) and to dilute highly enriched uranium from weapons stockpiles which is now being redirected to become reactor fuel. This dilution, also called downblending, means that any nation or group that acquired the finished fuel would have to repeat the (very expensive and complex) enrichment process before assembling a weapon.

Back end

The back end of the nuclear fuel cycle, mostly spent fuel rods, contains fission products that emit beta and gamma radiation, and actinides that emit alpha particles, such as uranium-234, neptunium-237, plutonium-238 and americium-241, and even sometimes some neutron emitters such as californium (Cf). These isotopes are formed in nuclear reactors.

It is important to distinguish the processing of uranium to make fuel from the reprocessing of used fuel. Used fuel contains the highly radioactive products of fission. Many of these are neutron absorbers, called neutron poisons in this context.

These eventually build up to a level where they absorb so many neutrons that the chain reaction stops, even with the control rods completely removed. At that point the fuel has to be replaced in the reactor with fresh fuel, even though there is still a substantial quantity of uranium-235 and plutonium present. In the United States, this used fuel is stored, while in countries such as the United Kingdom, France, and Japan, the fuel is reprocessed to remove the fission products, and the fuel can then be re-used. This reprocessing involves handling highly radioactive materials, and the fission products removed from the fuel are a concentrated form of high-level waste as are the chemicals used in the process.

Proliferation concerns

When dealing with uranium and plutonium, the possibility that they may be used to build nuclear weapons is often a concern. Active nuclear reactors and nuclear weapons stockpiles are very carefully safeguarded and controlled. However, high-level waste from nuclear reactors may contain plutonium.

Ordinarily, this plutonium is reactor-grade plutonium, containing a mixture of plutonium-239 (highly suitable for building nuclear weapons), plutonium-240 (an undesirable contaminant and highly radioactive), plutonium-241, and plutonium-238; these isotopes are difficult to separate. Moreover, high-level waste is full of highly radioactive fission products. However, most fission products are relatively short-lived. This is a concern since if the waste is stored, perhaps in deep geological storage, over many years the fission products decay, decreasing the radioactivity of the waste and making the plutonium easier to access.

Moreover, the undesirable contaminant Pu-240 decays faster than the Pu-239, and thus the quality of the bomb material increases with time (although its quantity decreases during that time as well). Thus, some have argued, as time passes, these deep storage areas have the potential to become "plutonium mines", from which material for nuclear weapons can be acquired with relatively little difficulty. Critics of the latter idea point out that the half-life of Pu-240 is 6,560 years and Pu-239 is 24,110 years, and thus the relative enrichment of one isotope to the other with time occurs with a half-life of 9,000 years (that is, it takes 9000 years for the *fraction* of Pu-240 in a sample of mixed plutonium isotopes, to spontaneously decrease by half—a typical enrichment needed to turn reactor-grade into weapons-grade Pu).

Thus "weapons grade plutonium mines" would be a problem for the very far future (>9,000 years from now), so that there remains a great deal of time for technology to advance to solve this problem, before it becomes acute.

Pu-239 decays to U-235 which is suitable for weapons and which has a very long half life (roughly 10^9 years). Thus plutonium may decay and leave uranium-235. However, modern reactors are only moderately enriched with U-235 relative to U-238, so the U-238 continues to serve as denaturation agent for any U-235 produced by plutonium decay.

One solution to this problem is to recycle the plutonium and use it as a fuel e.g. in fast reactors. But in the minds of some, the very existence of the nuclear fuel reprocessing plant needed to separate the plutonium from the other elements represents a proliferation concern. In pyrometallurgical fast reactors, the waste generated is an actinide compound that cannot be used for nuclear weapons.

Goals of waste management

The main objective in managing and disposing of radioactive (or other) waste is to protect people and the environment. This means isolating or diluting the waste so that the rate or concentration of any radionuclide returned to the biosphere is harmless. To achieve this the preferred technology to date has been deep and secure burial for the more dangerous wastes; transmutation, long-term retrievable storage, and removal to space have also been suggested.

Radioactivity by definition reduces over time, so in principle the waste needs to be isolated for a particular period of time until its components have decayed such that it no longer poses a threat. In practice this can mean periods of hundreds of thousands of years, depending on the nature of the waste involved. Though an affirmative answer is often taken for granted, the question as to whether or not we should endeavor to avoid causing harm to remote future generations, perhaps thousands upon thousands of years hence, is essentially one which must be dealt with by philosophy.

Sources of waste

Radioactive waste comes from a number of sources. The majority of waste originates from the nuclear fuel cycle and nuclear weapons reprocessing. However, other sources include medical and industrial wastes, as well as naturally occurring radioactive materials (NORM) that can be concentrated as a result of the processing or consumption of coal, oil and gas, and some minerals.

HAZARDOUS WASTE

Any solid, liquid, or gaseous waste materials that, if improperly managed or disposed of, may pose substantial hazards to human health and the environment. Every industrial country in the world has had problems with managing hazardous wastes. Improper disposal of these waste streams in the past has created a need for very expensive cleanup operations. Efforts are under way internationally to remedy old problems caused by hazardous waste and to prevent the occurrence of other problems in the future. A waste is considered hazardous if it exhibits one or more of the following characteristics: ignitability, corrosivity, reactivity, and toxicity. Ignitable wastes can create fires under certain conditions; examples include liquids, such as solvents, that readily catch fire, and friction-sensitive substances. Corrosive wastes include those that are acidic and those that are capable of corroding metal (such as tanks, containers, drums, and barrels). Reactive wastes are unstable under normal conditions. They can create explosions, toxic fumes, gases, or vapors when mixed with water. Toxic wastes are harmful or fatal when ingested or absorbed. When they are disposed of on land, contaminated liquid may drain (leach) from the waste and pollute groundwater.

Put simply, a hazardous waste is waste that poses substantial or potential threats to public health or the environment and generally exhibits one or more of these characteristics:

- ignitable (i.e., flammable)
- oxidizing
- corrosive
- toxic
- radioactive

U.S. environmental laws additionally describe a "hazardous waste" as a waste (usually a solid waste) that has the potential to:

- cause, or significantly contribute to an increase in mortality (death) or an increase in serious irreversible, or incapacitating reversible illness; or

- pose a substantial present or potential hazard to human health or the environment when improperly treated, stored, transported, or disposed of, or otherwise managed.

The term "hazardous waste" comprises all toxic chemicals, radioactive materials, and biologic or infectious waste. These materials threaten workers through occupational exposure and the general public in their homes, communities, and general environment.

Exposure to these materials can occur near the site of generation, along the path of its transportation, and near their ultimate disposal sites. Most hazardous waste results from industrial processes that yield unwanted people that use byproducts, defective products, and spilled materials. The generation and disposal of hazardous wastes is controlled through a variety of international and national regulations. Hazardous waste was formerly known as 'special' waste.

Resource Conservation and Recovery Act (RCRA)

Modern hazardous waste regulations in the U.S. began with the Resource Conservation and Recovery Act (RCRA) which was enacted in 1976. The primary contribution of RCRA was to create a "cradle to grave" system of record keeping for hazardous wastes. Hazardous wastes must be tracked from the time they are generated until their final disposition.

RCRA's record keeping system helps to track the life cycle of hazardous waste and reduces the amount of hazardous waste illegally disposed.

Many types of businesses generate hazardous waste. Some are small areas that may be located in a community. For example, dry cleaners, automobile repair shops, hospitals, exterminators, and photo processing centers all generate hazardous waste. Some hazardous waste generators are larger companies like chemical manufacturers, electroplating companies, and oil refineries.

A US facility that treats, stores or disposes of hazardous waste must obtain a permit for doing so under the Resource Conservation and Recovery Act. Generators of and transporters of hazardous waste must meet specific requirements for handling, managing, and tracking waste. Through the RCRA, Congress directed the United States Environmental Protection Agency (EPA) to create regulations to manage hazardous waste. Under this mandate, the EPA developed strict requirements for all aspects of hazardous waste management including the treatment, storage, and disposal of hazardous waste. In addition to these federal requirements, states may develop more stringent requirements or requirements that are broader in scope than the federal regulations. In the United States, hazardous wastes generated by commercial or industrial activities may be classified as "listed" hazardous wastes or "characteristic" hazardous wastes by the EPA.

In regulatory terms, a Resource Conservation and Recovery Act (RCRA) hazardous waste is a waste that either a "characteristic waste" or a "listed waste":

- Characteristic Waste - exhibits at least one of the four "characteristics" of hazardous waste (ignitability, corrosivity, reactivity, or toxicity)
- Listed Waste - appears on one of the four hazardous wastes lists (F-list, K-list, P-list, or U-list), or

Individual states may regulate particular wastes more stringently than mandated by federal regulation. This is because the U.S. EPA is authorized to delegate primary rulemaking authorization to individual states. Most states take advantage of this authority and form their own state regulatory agencies that are monitored by the U.S. EPA.

Characteristic Wastes

Characteristic Hazardous Wastes are defined as wastes that exhibit the following characteristics: ignitability, corrosivity, reactivity, or toxicity.

Toxicity Characteristic Leaching Procedure (TCLP)

Characteristic wastes are identified in the U.S. through a laboratory procedure called the toxicity characteristic leaching procedure (TCLP). The TCLP helps identify wastes likely to leach concentrations of contaminants into the environment that may be harmful to human health or the environment.

Exempted hazardous wastes

USEPA regulations automatically exempt certain solid wastes from being regulated as "hazardous wastes". This does not necessarily mean the wastes are not hazardous nor that they are not regulated.

An exempted hazardous waste simply means that the waste is not regulated by the primary hazardous waste regulations. Many of these wastes may by regulated by different statutes and/or regulations and/or by different regulatory agencies. For example, many hazardous mining wastes are regulated via mining statutes and regulations. "Exempted" hazardous wastes include:

- Household hazardous waste (HHW);
- Agricultural wastes which are returned to the ground as fertilizer;
- Mining overburden returned to the mine site;
- Utility wastes from [coal] combustion to produce electricity;
- Oil and natural gas exploration drilling waste;
- Wastes from the extraction of beneficiation, and processing of ores and minerals, including coal;

- Cement kiln wastes;
- Wood treated with arsenic preservatives.
- Certain chromium-containing wastes
- Recycled hazardous wastes: Some hazardous wastes that are recycled may also be exempted from hazardous waste regulations.

Household Hazardous Waste

Household Hazardous Waste (HHW) (also referred to as domestic hazardous waste) is waste that is generated from residential households. HHW only applies to wastes that are the result of the use of materials that are labeled for and sold for "home use".

The following list includs categories often applied to HHW. It is important to note that many of these categories overlap and that many household wastes can fall into multiple categories:

- Paints and solvents
- Automotive wastes (used motor oil, antifreeze, etc.)
- Pesticides (insecticides, herbicides, fungicides, etc.)
- Mercury-containing wastes (thermometers, switches, fluorescent lighting, etc)
- Electronics (computers, televisions, cell phones)
- Aerosols / Propane cylinders
- Caustics / Cleaning agents
- Refrigerant-containing appliances
- Batteries
- Ammunition
- Radioactive waste (some home smoke detectors are classified as radioactive waste because they contain very small amounts of a radioactive isotope of americium.

Disposal of HHW in the United States

Because of the expense associated with the disposal of HHW, it is still legal for most homeowners in the U.S. to dispose of most types of household hazardous wastes as municipal solid waste (MSW) and these wastes can be put in our trash.

Laws vary by state and municipality and they are changing every day. Be sure to check with our local environmental regulatory agency, solid waste authority, or health department to find out how HHW is managed in our area. Modern landfills are designed to handle normal amounts of HHW and minimize the environmental impacts. However, there are still going to be some impacts and there are many ways that homeowners can keep these wastes out of landfills.

Laws regulating HHW in the U.S. are gradually becoming more strict. As of 2007, radioactive smoke detectors are the only HHW that are managed nationally.

While it is still legal in the United States to dispose of smoke detectors in our trash in most places, manufacturers of smoke detectors must accept returned units for disposal as mandated by the Nuclear Regulatory law 10 CFR 32.27. If we send our detector back to a manufacturer then it will be disposed in a nuclear waste facility.

In the U.S., states are regulating various HHW waste disposal in MSW landfills on a state by state basis. Some commonly regulated wastes in some (but not all) states include restrictions on the disposal of:

- Recyclables (especially "source-separated" recyclables or recyclables that have already been separated from solid waste). In this case this would only apply to household hazardous wastes that have been separated for recycling.
- Lead-acid batteries
- Mercury-containing wastes
- Rechargeable batteries
- Cathode ray tubes (CRTs) from older computer monitors and televisions
- Cell phones and computers
- Refrigerant containing appliances such as a refrigerator, air conditioner or dehumidifier.

(Note: Yard waste or "green waste" (particularly "source-separated" yard waste such as from a city leaf collection program) is not hazardous but may be a regulated household waste)

Local solid waste authorities and health departments may also have specific bans on wastes that apply to their service area.

Listed Wastes

Listed hazardous wastes are generated by specific industries and processes and are automatically considered hazardous, based solely on the process that generates them and irrespective of whether a test of the waste shows any of the "characteristics" of hazardous waste. Examples of listed wastes include:

- many sludges leftover from electroplating processes.
- certain waste from iron and steel manufacturing
- wastes from certain cleaning and/or degreasing processes

Hazardous wastes are incorporated into lists published by the Environmental Protection Agency. These lists are organized into three categories:

The F-list (Non-specific Source Wastes)

non-specific This list identifies wastes from common manufacturing and industrial processes, such as solvents, that have been used in cleaning or degreasing operations. Because the processes producing these wastes can occur in different sectors of industry, the F-listed wastes are known as wastes from sources.

The K-list (Source-specific Wastes)

This list includes certain wastes from specific industries, such as petroleum refining or pesticide manufacturing. Certain sludges and wastewaters from treatment and production processes in these industries are examples of source-specific wastes.

Discarded Wastes (P-List and U-List)

P-List and U-List wastes are actually sublists of the same major list applying to discarded wastes. These wastes apply to commercial chemical products that are considered hazardous when discarded and are regulated under the following U.S. Federal Regulation: 40 C.F.R. 261.33(e) and 261.33(f). P-List wastes are wastes that are considered "acutely hazardous" when discarded and are subject to more stringent regulation. Nitric oxide is an example of a P-list waste and carries the number P076. U-Listed wastes are considered "hazardous" when discarded and are regulated in a somewhat less stringent manner than P-Listed wastes.

Universal Wastes

Universal wastes are hazardous wastes that (in the U.S.):

- generally pose a lower threat relative to other hazardous wastes
- are ubiquitous and produced in very large quantities by a large number of generators.

Some of the most common "universal wastes" are: fluorescent light bulbs, batteries, cathode ray tubes, and mercury-containing devices.

Universal wastes are subject to somewhat less sringent regulatory requirements and small quantity generators of universal wastes may be classified as "conditionally-exempt small quantity generators" (CESQGs) which releases them from some of the regulatory requirements for the handling and storage hazardous wastes.

Solid Waste Haulers and HHW

One "catch-22" that residents often encounter is that while it may be legal to dispose of some HHW in their regular trash, the waste hauler that collects the trash can choose not to haul the waste. It is not uncommon for a waste hauler to refuse to pick up municipal solid waste that contains things like paint and fluorescent light bulbs. There is often little recourse for residents in this case. In these cases the resident may have to make their own arrangements to dispose of the waste by taking it directly to a landfill or solid waste transfer station.

Final disposition of hazardous waste

Hazardous wastes (HWs) are typically dealt with in five different ways:

Recycling

Many HWs can be recycled into new products. Examples might include lead-acid batteries or electronic circuit boards where the heavy metals can be recovered and used in new products.

Neutralization

Some HW can be processed so that the hazardous component of the waste is eliminated. making it a non-hazardous waste. An example of this might include a corrosive acid that is neutralized with a basic substance so that it is no-longer corrosive.

Incineration, Destruction and Waste-to-energy

A HW may be "destroyed" for example by incinerating it at a high temperature. Flammable wastes can sometimes be burned as energy sources. For example many cement kilns burn HWs like used oils or solvents.

Hazardous waste landfill (sequestering, isolation, etc.)

A HW may be sequestered in a HW landfill or permanent disposal facility. "In terms of hazardous waste, a landfill is defined as a disposal facility or part of a facility where hazardous waste is placed in or on land and which is not a pile, a land treatment facility, a surface impoundment, an underground injection well, a salt dome formation, a salt bed formation, an underground mine, a cave, or a corrective action management unit.

ENVIRONMENTAL REMEDIATION

Environmental remediation involves providing a remedy for an environmental problem. This can include removing contaminants from groundwater or cleaning up after an oil spill. Remediation is not always just subject to the will of the people, but is often a matter of government regulation or intervention.

One of the main purposes of environmental remediation is to restore contaminated sites or resources to a level that is safe for humans and animals. Depending on the type of damage that is done, this can be a complex and expensive process. There are companies that specialize in environmental remediation. Even with the help of these professionals and environmental experts, however, sometimes there is nothing that can be done to restore a contaminated site to a point where it is safe.

There are various remediation technologies employed around the world. Since environmental remediation is often more costly if contamination occurs, many companies try to take preventative measures. They employ relevant technology to treat their industrial waste before it comes into contact with the environment. Examples of these methods include incineration, soil washing, and chemical precipitation.

Environmental remediation tactics are sometimes divided into groups called ex-situ and in-situ. Processes that involve excavation of soil are considered ex-situ. In-situ procedures are those that attempt to treat contamination without removing soil.

Remediation Standards

In the USA the most comprehensive set of Preliminary Remediation Goals (PRGs) is from the Environmental Protection Agency (EPA) Region 9. A set of standards used in Europe exists and is often called the Dutch standards. The European Union (EU) is rapidly moving towards Europe-wide standards, although most of the industrialised nations in Europe have their own standards at present.

Funding Remediation

In the US there has been a mechanism for taxing polluting industries to form a Superfund to remediate abandoned sites, or to litigate to force corporations to remediate their contaminated sites. Other countries have other mechanisms and commonly sites are rezoned to "higher" uses such as high density housing, to give the land a higher value so that after deducting clean up costs there is still an incentive for a developer to purchase the land, clean it up, redevelop it and sell it on, often as apartments (home units).

Remediation Technologies

Remediation technologies are many and varied but can be categorised into ex-situ and in-situ methods. Ex-situ methods involve excavation of impacted soils and subsequent treatment at the surface, In-situ methods seek to treat the contamination without removing the soils.

The more traditional remediation approach (used almost exclusively on contaminated sites from the 1970s to the 1990s) consists primarily of soil excavation and disposal to landfill "dig and dump" and groundwater "pump and treat".

Site Assessment

Once a site is suspected of being contaminated there is a need to assess the contamination. Often the assessment begins with preparation of a Phase I Environmental Site Assessment. The historical use of the site and the materials used and produced on site will guide the assessment strategy and type of sampling and chemical analysis to be done. Often nearby sites owned by the same company or which are nearby and have been reclaimed, levelled or filled are also contaminated even where the current land use seems innocuous. For example, a car park may have been levelled by using contaminated waste in the fill. Also important is to consider off site contamination of nearby sites often through decades of emissions to soil, groundwater, and air. Ceiling dust, topsoil, surface and groundwater of nearby properties should also be tested, both before and after any remediation. This is a controversial step as:

- No one wants to have to pay for the clean up of the site;
- If nearby properties are found to be contaminated it may have to be noted on their property title, potentially affecting the value;
- No one wants to pay for the cost of assessment.

Often corporations which do voluntary testing of their sites are protected from the reports to environmental agencies becoming public under Freedom of Information Acts, however a "Freedom of Information" inquiry will often produce other documents that are not protected or will produce references to the reports.

Excavation or dredging

Excavation processes can be as simple as hauling the contaminated soil to a regulated landfill, but can also involve aerating the excavated material in the case of volatile organic compounds (VOCs). If the contamination affects a river or bay bottom, then dredging of bay mud or other silty clays containing contaminants may be conducted.

SEAR - surfactant enhanced aquifer remediation

Also know as Solubilization and recovery, the *Surfactant Enhanced Aquifer Remediation* process involves the injection of hydrocarbon mitigation agents or specialty surfactants into the subsurface to enhance desorption and recovery of bound up otherwise recalcitrant non aqueous phase liquid (NAPL).

In geologic formations that allow delivery of hydrocarbon mitigation agents or specialty surfactants, this approach provides a cost effective and permanent solution to sites that have been previously unsuccessful utilizing other remedial approaches. This technology is also successful when utilized as the initial step in a multi faceted remedial approach utilizing SEAR then In situ Oxidation, bioremediation enhancement or soil vapor extraction (SVE).

In situ oxidation

New *in situ oxidation* technologies have become popular, for remediation of a wide range of soil and groundwater contaminants. Remediation by chemical oxidation involves the injection of strong oxidants such as hydrogen peroxide, ozone gas, potassium permanganate or persulfates. Oxygen gas or ambient air can also be injected as a more mild approach. One disadvantage of this approach is the possibility of less contaminant destruction by natural attenuation if the bacteria which normally live in the soil prefer a reducing environment. The injection of gases into the groundwater may also cause contamination to spread faster than normal depending on the site's hydrogeology.

Soil vapor extraction

Soil vapor extraction (SVE) is an effective remediation technology for soil. "Multi Phase Extraction" (MPE) is also an effective remediation technology when soil and groundwater are to be remediated coincidentally. SVE and MPE utilize different technologies to treat the off-gas volatile organic compounds (VOCs) generated after vacuum removal of air and vapors (and VOCs) from the subsurface and include granular activated carbon (most commonly used historically), thermal and/or catalytic oxidation and vapor condensation. Generally, carbon is used for low (<500ppmV) VOC concentration vapor streams, oxidation is used for moderate (up to 4,000 ppmV) VOC concentration streams, and vapor condensation is used for high (>4,000 ppmV) VOC concentration vapor streams. Granular activated carbon (GAC) is used as a filter for air or water. Commonly used to filter tap water in household sinks. GAC is a highly porous adsorbent material, produced by heating organic matter, such as coal, wood and coconut shell, in the absence of air, which is then crushed into granules. Activated carbon is positively charged and therefore able to remove negative ions from the water such as organic ions, ozone, chlorine, fluorides and dissolved organic solutes by adsorption onto the activated carbon. The activated carbon must be replaced periodically as it may become saturated and unable to adsorb (i.e. reduced absorption efficiency with loading). Activated carbon is not effective in removing heavy metals.

Thermal oxidation (or incineration) can also be an effective remediation technology. This approach is somewhat controversial because of the risks of dioxins released in the atmosphere through the exhaust gases or effluent off-gas. Controlled, high temperature incineration with filtering of exhaust gases however should not pose any risks. Two different technologies can be employed to oxidize the contaminants of an extracted vapor stream. The selection of either thermal or catalytic depends on the type and concentration in parts per million by volume of consituent in the vapor stream. Thermal oxidation is more useful for higher concentration (~4,000 ppmV) influent vapor streams (which require less natural gas usage) than catalytic oxidation at ~2,000 ppmV.

- Thermal oxidation which uses a system that acts as a furnace and maintains temperatures ranging from 1350°F to 1500°F (730°C-815°C).
- Catalytic oxidation which uses a catalyst on a support to facilitate a lower temperature oxidation. This system usually maintains temperatures ranging from 600°F to 800°F (315°C-430°C).
- Vapor condensation is the most effective off-gas treatment technology for high (>4,000 ppmV) VOC concentration vapor streams. The process involves cryogenically cooling the vapor stream to below 40 degrees C such that the VOCs condensate out of the vapor stream and into liquid form where it is collected in steel containers. The liquid form of the VOCs is referred to as dense non-aqueous phase liquids (DNAPL) when the source of the liquid consists predominantly of solvents or light non-aqueous phase liquids (LNAPL) when the source of the liquid consists predominantly of petroleum or fuel products. This recovered chemical can then be reused or recycled in a more environmentally sustainable or green manner than the alternatives. This technology is also known as cryogenic cooling and compression (C3-Technology).

Pump and Treat

Pump and treat involves pumping out contaminated groundwater with the use of a submersible or vacuum pump, and allowing the extracted groundwater to be purified by slowly proceeding through a series of vessels that contain materials designed to adsorb the contaminants from the groundwater.

For petroleum-contaminated sites this material is usually activated carbon in granular form. Chemical reagents such as flocculants followed by sand filters may also be used to decrease the contamination of groundwater. Air stripping is a method that can be effective for volatile pollutants such as BTEX compounds found in gasoline.

For most biodegradable materials like BTEX, MTBE and most hydrocarbons, bioreactors can be used to clean the contaminated water to non-detectable levels. With fluidized bed bioreactors it is possible achieve very low discharge concentrations which will meet or exceed discharge standards for most pollutants.

Depending on geology and soil type, pump and treat may be a good method to quickly reduce high concentrations of pollutants. It is more difficult to reach sufficiently low concentrations to satisfy remediation standards, due to the equilibrium of absorption (chemistry)/desorption processes in the soil.

Other Technologies

The treatment of environmental problems through biological means is known as *bioremediation* and the specific use of plants for example by using phytoremediation. Bioremediation is sometimes used in conjunction with a pump and treat system. In bioremediation, either naturally occurring or specially bred bacteria are used to consume contaminants from extracted groundwater. This is sometimes referred to as a bio-gac system. Many times the groundwater is recycled to allow for continuously flowing water and enhanced bacteria population growth. Occasionally the bacteria can build up to such a point that they can affect filtration and pumping. The vessel should then be partially drained. Care must be taken to ensure that a sharp change in the groundwater chemistry does not kill the bacteria (such as a sudden change in pH). Dual-phase extraction utilizes a soil vapor extraction system that produces a high vacuum resulting in the extraction of both contaminated vapors as well as a limited amount of contaminated groundwater. This method is somewhat inefficient due to large amount of energy required by pulling water by vacuum compared to pushing water with a submersible pump.

Emissions Standards

Standards are set for the levels of dust, noise, odour, emissions to air and groundwater, and discharge to sewers or waterways of all chemicals of concern or chemicals likely to be produced during the remediation by processing of the contaminants. These are compared against both natural background levels in the area and standards for areas zoned as nearby areas are zoned and against standards used in other recent remediations. Just because the emission is emanating from an area zoned industrial does not mean that in a nearby residential area there should be permitted any exceedances of the appropriate residential standards. Monitoring for compliance against each standards is critical to ensure that exceedances are detected and reported both to authorities and the local community. Enforcement is necessary to ensure that continued or significant breeches result in fines or even a jail sentence for the polluter. Penalties must be significant as otherwise fines are treated as a normal expense of doing business. Compliance must be cheaper than to have continuous breeches.

Transport and Emergency Safety Assessment

Assessment should be made of the risks of operations, transporting contaminated material, disposal of waste which may be contaminated including workers' clothes, and a formal emergency response plan should be developed. Every worker and visitor entering the site should have a safety induction personalised to their involvement with the site.

Impacts of Funding Remediation

The rezoning is often resisted by local communities and local government because of the adverse impacts on the local amenity of the remediation and the new development. The main impacts during remediation are noise, dust, odour and incremental health risk. Then there is the noise, dust and traffic of developments. Then there is the impact on local traffic, schools, playing fields, and other public facilities of the often vastly increased local population.

This rezoning, remediation and redevelopment of land contaminated by Union Carbide, ICI and others also involves the remediation of a strip of dioxin contaminated sediments in Homebush Bay, New South Wales. The Homebush Bay area was home to the main events of the Sydney 2000 Summer Olympics. The soil contamination was addressed in the Commission of Inquiry into the Lednez site formerly owned by Union Carbide, but not to the satisfaction of local community activists.

The remediation of Homebush Bay is important because of its impact on the food chain which extends through benthos not only to local protected and threatened species of birds, but also to JAMBA and CAMBA protected species and species which use other Ramsar-protected wetlands. Ultimately human health is impacted through the food chain. Homebush Bay has a complete fishing ban, there is a commercial fin fishing ban west of the Gladesville Bridge, and based on submissions of the remediator and NSW Waterways and EPA the complete fishing ban ought be extended to the whole of the Parramatta River west of Homebush Bay and at least as far East as the Ryde Traffic Bridge.

Community Consultation and Information

In preparation for any significant remediation there should be extensive community consultation. The proponent should both present information to and seek information from the community. The proponent needs to learn about "sensitive" (future) uses like childcare, schools, hospitals, and playgrounds as well as community concerns and interests information.

Consultation should be open, on a group basis so that each member of the community is informed about issues they may not have individually thought about. An independent chairperson acceptable to both the proponent and the community should be engaged (at proponent expense if a fee is required). Minutes of meetings including questions asked and the answers to them and copies of presentations by the proponent should be available both on the internet and at a local library (even a school library) or community centre.

6

Factors Affecting Pesticide Toxicity in Aquatic Systems

The ecological impacts of pesticides in water are determined by the following criteria:

Toxicity: Mammalian and non-mammalian toxicity usually expressed as LD_{50} ("Lethal Dose": concentration of the pesticide which will kill half the test organisms over a specified test period). The lower the LD_{50}, the greater the toxicity; values of 0-10 are extremely toxic.

Drinking water and food guidelines are determined using a risk-based assessment. Generally, Risk = Exposure (amount and/or duration) × Toxicity.

Toxic response (effect) can be acute (death) or chronic (an effect that does not cause death over the test period but which causes observable effects in the test organism such as cancers and tumours, reproductive failure, growth inhibition, teratogenic effects, etc.).

Persistence: Measured as half-life (time required for the ambient concentration to decrease by 50 per cent). Persistence is determined by biotic and abiotic degradational processes. Biotic processes are biodegradation and metabolism; abiotic processes are mainly hydrolysis, photolysis, and oxidation. Modern pesticides tend to have short half lives that reflect the period over which the pest needs to be controlled.

Degradates:

The degradational process may lead to formation of "degradates" which may have greater, equal or lesser toxicity than the parent compound. As an example, DDT degrades to DDD and DDE.

Fate (Environmental): The environmental fate (behaviour) of a pesticide is affected by the natural affinity of the chemical for one of four environmental compartments: solid matter (mineral matter and particulate organic carbon), liquid (solubility in surface and soil water), gaseous form (volatilization), and biota.

This behaviour is often referred to as "partitioning" and involves, respectively, the determination of: the soil sorption coefficient (K_{OC}); solubility; Henry's Constant (H); and the n-octanol/water partition coefficient (K_{OW}). These parameters are well known for pesticides and are used to predict the environmental fate of the pesticide.

An additional factor can be the presence of impurities in the pesticide formulation but that are not part of the active ingredient. A recent example is the case of TFM, a lampricide used in tributaries of the Great Lakes for many years for the control of the sea lamprey. Although the environmental fate of TFM has been well known for many years, recent research by Munkittrick *et al.* has found that TFM formulation includes one or more highly potent impurities that impact on the hormonal system of fish and cause liver disease.

Ecological Effects of Pesticides

Pesticides are included in a broad range of organic micro pollutants that have ecological impacts. Different categories of pesticides have different types of effects on living organisms, therefore generalization is difficult. Although terrestrial impacts by pesticides do occur, the principal pathway that causes ecological impacts is that of water contaminated by pesticide runoff. The two principal mechanisms are bioconcentration and biomagnification.

Bioconcentration

This is the movement of a chemical from the surrounding medium into an organism. The primary "sink" for some pesticides is fatty tissue ("lipids"). Some pesticides, such as DDT, are "lipophilic", meaning that they are soluble in, and accumulate in, fatty tissue such as edible fish tissue and human fatty tissue. Other pesticides such as glyphosate are metabolized and excreted.

Biomagnification

This term describes the increasing concentration of a chemical as food energy is transformed within the food chain. As smaller organisms are eaten by larger organisms, the concentration of pesticides and other chemicals are increasingly magnified in tissue and other organs. Very high concentrations can be observed in top predators, including man.

The ecological effects of pesticides (and other organic contaminants) are varied and are often inter-related. Effects at the organism or ecological level are usually considered to be an early warning indicator of potential human health impacts. The major types of effects are listed below and will vary depending on the organism under investigation and the type of pesticide. Different pesticides have markedly different effects on aquatic life which makes generalization very difficult. The important point is that many of these effects are chronic (not lethal), are often not noticed by casual observers, yet have consequences for the entire food chain.

- Death of the organism.
- Cancers, tumours and lesions on fish and animals.
- Reproductive inhibition or failure.
- Suppression of immune system.
- Disruption of endocrine (hormonal) system.
- Cellular and DNA damage.
- Teratogenic effects (physical deformities such as hooked beaks on birds).
- Poor fish health marked by low red to white blood cell ratio, excessive slime on fish scales and gills, etc.
- Intergenerational effects (effects are not apparent until subsequent generations of the organism).
- Other physiological effects such as egg shell thinning.

These effects are not necessarily caused solely by exposure to pesticides or other organic contaminants, but may be associated with a combination of environmental stresses such as eutrophication and pathogens. These associated stresses need not be large to have a synergistic effect with organic micro pollutants. Ecological effects of pesticides extend beyond individual organisms and can extend to ecosystems. Swedish work indicates that application of pesticides is thought to be one of the most significant factors affecting biodiversity. Jonsson *et al.* report that the continued decline of the Swedish partridge population is linked to changes in land use and the use of chemical weed control.

Chemical weed control has the effect of reducing habitat, decreasing the number of weed species, and of shifting the balance of species in the plant community. Swedish studies also show the impact of pesticides on soil fertility, including inhibition of nitrification with concomitant reduced uptake of nitrogen by plants. These studies also suggest that pesticides adversely affect soil micro-organisms which are responsible for microbial degradation of plant matter (and of some pesticides), and for soil structure.

MONITORING DATA FOR PESTICIDES

Monitoring data for pesticides are generally poor in much of the world and especially in developing countries. Key pesticides are included in the monitoring schedule of most western countries, however the cost of analysis and the necessity to sample at critical times of the year (linked to periods of pesticide use) often preclude development of an extensive data set.

Many developing countries have difficulty carrying out organic chemical analysis due to problems of inadequate facilities, impure reagents, and financial constraints. New techniques using immunoassay procedures for presence/absence of specific pesticides may reduce costs and increase reliability. Immunoassay tests are available for triazines, acid amides, carbamates, 2,4-D/phenoxy acid, paraquot and aldrin.

Data on pesticide residues in fish for lipophilic compounds, and determination of exposure and/or impact of fish to lipophobic pesticides through liver and/or bile analysis is mainly restricted to research programmes.

Hence, it is often difficult to determine the presence, pathways and fate of the range of pesticides that are now used in large parts of the world.

In contrast, the ecosystemic impacts from older, organochlorine pesticides such as DDT, became readily apparent and has resulted in the banning of these compounds in many parts of the world for agricultural purposes.

Older pesticides, together with other hydrophobic carcinogens such as PAHs and PCBs, are poorly monitored when using water samples.

As an example, the range of concentration of suspended solids in rivers is often between 100 and 1000 mg/l except during major runoff events when concentrations can greatly exceed these values. Tropical rivers that are unimpacted by development have very low suspended sediment concentrations, but increasingly these are a rarity due to agricultural expansion and deforestation in tropical countries. As an example, approximately 67 per cent of DDT is transported in association with suspended matter at sediment concentrations as low as 100 mg/l, and increases to 93 per cent at 1000 mg/l of suspended sediment.

PESTICIDES AND TOXIC CHEMICALS

Pesticides typically enter a waterbody through surface water runoff, often from a farm field or from neighborhoods where they are applied on lawns. Pesticides can also enter a waterbody as a result of "spray drift." This occurs when the pesticide is sprayed over an area, and the wind blows some of the spray into a nearby waterbody.

Pesticides are designed to be toxic to a target organism, but they often kill other organisms as well. The insecticide azinphos-methyl, for example, is used to control insects such as biting mites and aphids. It is also very toxic to fish and birds, however. For the most part, today's pesticides do not build up in the tissues of animals -- a process called bioaccumulation -- to the extent that older compounds like DDT did. On the other hand, many of the compounds used today are toxic at very low concentrations.

Toxic chemicals, such as spilled oils and fuels in cities, are often washed off streets, down storm drains, and into waterbodies. Combustion of fuels in automobiles and factories introduces hydrocarbons and metals into the environment. They eventually end up in the water through atmospheric deposition or runoff. Industrial facilities without the proper means to control runoff can also contribute toxic chemicals to the aquatic environment. The type of chemical that is released depends on the type of manufacturing done at a facility. Other chemicals, such as solvents, paints, cleaning solutions and others, originate from marinas and boating activities.

Persitant Pollutants/Pesticides

Environmentally-persistent pollutants originate from agricultural and urban pesticide use, combustion, and industrial production and can accumulate in sediments, soils, vegetation, animals, and humans. Of particular concern are the persistent highly lipid-soluble, semi-volatile organic compounds (SVOCs). Despite growing concerns, environmental science has failed to provide a reliable regional-scale framework for characterizing concentrations, dispersion and human exposure for persistent SVOCs. Efforts to move from an existing qualitative to a more quantitative exposure assessment are hindered by a number of scientific obstacles. One problem is the accuracy in measurements of the partitioning of SVOCs, between air and airborne particulate matter, between air and soils, and between air and vegetation--each a complex intermedia transfer. The goal of this research component is to develop a multi-domain intermedia transport analysis that can capture space and time scales for the coupled domain transport of some persistent pollutants. This research component is also directed to selection of the most appropriate time scale for an exposure assessment, based on both pharmacokinetics and on the biokinetics of the health effects for a given pollutant. This will make possible the definition of appropriate time scales for both the measurement and modeling of human exposure to multimedia pollutants.

We are developing the methods and findings for this research component through eight areas of research:

- determination of chemical/emission-specific characteristic space and time scales,
- development of an imbedded multi-scale approach for multimedia transport and transformation modeling,
- the use of pharmacokinetic and toxicokinetic models to link intake and uptake to measures of dose,
- development of a multipathway exposure model that can be linked with the multi-scale, multimedia transport and transformation model,

- development and application of calibration/validation methods for multimedia source-to-dose models,
- assessing chemical biotransfer among air, soil and vegetation,
- intermedia transfer factors from chemicals on air particles to human intake, and
- uncertainty and sensitivity analysis methods.

AQUACULTURE AND AGRICULTURAL POLLUTION

Aquaculture is a special case of agricultural pollution. There are two main forms: land-based and water-based systems. Effluent controls are possible on land-based systems, however water-based systems present particular problems. Aquaculture is rapidly expanding in most parts of the developed and developing world, both in freshwater and marine environments. In contrast, coastal fisheries in most countries are declining.

The environmental impact is primarily a function of feed composition and feed conversion (faecal wastes), plus assorted chemicals used as biocides, disinfectants, medicines, etc. Wastage of feed (feed not taken up by the fish) is estimated to be 20 per cent in European aquaculture. Waste feed and faecal production both add substantial nutrient loadings to aquatic systems.

Additional environmental problems include risk of disease and disease transfer to wild fish, introduction of exotic species, impacts on benthic communities and on the eutrophication of water, interbreeding of escaped cultured fish with wild fish with consequent genetic change in the wild population. Traditional integrated aquaculture systems, as in China, where sewage-fish culture is practised, can be a stabilizing influence in the entire ecosystem. This is recommended, especially in developing countries where water and resources are scarce or expensive.

Problems of Restoration of Eutrophic Lakes

Eutrophic and hypertrophic lakes tend to be shallow and suffer from high rates of nutrient loadings from point and non-point sources. In areas of rich soils such as the Canadian prairies, lake bottom sediments are comprised of nutrient-enriched soil particles eroded from surrounding soils. The association of phosphorus with sediment is a serious problem in the restoration of shallow, enriched lakes. P-enriched particles settle to the bottom of the lake and form a large pool of nutrient in the bottom sediments that is readily available to rooted plants and which is released from bottom sediments under conditions of anoxia into the overlying water column and which is quickly utilized by algae.

This phosphorus pool, known as the "internal load" of phosphorus, can greatly offset any measures taken by river basin managers to control lake eutrophication by control of external phosphorus sources from agriculture and from point sources. Historically, dredging of bottom sediments was considered the only means of remediating nutrient-rich lake sediments, however, modern technology now provides alternative and more cost-effective methods of controlling internal loads of phosphorus by oxygenation and by chemically treating sediments *in situ* to immobilize the phosphorus. Nevertheless, lake restoration is expensive and must be part of a comprehensive river basin management programme.

SEDIMENTS POLLUTION

Sediments in the aquatic ecosystem are analogous to soil in the terrestrial ecosystem as they are the source of substrate nutrients, and micro- and macroflora and -fauna that are the basis of support to living aquatic resources. Sediments are the key catalysts of environmental food cycles and the dynamics of water quality. Aquatic sediments are derived from and composed of natural physical, chemical, and biological components generally related to their watersheds.

Global estimates of erosion and sediment transport in major rivers of the world vary widely, reflecting the difficulty in obtaining reliable values for sediment concentration and discharge in many countries, the assumptions that are made by different researchers, and the opposing effects of accelerated erosion due to human activities (deforestation, poor agricultural practices, road construction, etc.) relative to sediment storage by dam construction. Milliman and Syvitski (1992) estimate global sediment load to oceans in the mid-20th century at 20 thousand million t/yr, of which about 30% comes from rivers of southern Asia (including the Yangtze and Yellow Rivers of China). Significantly, they believe that almost 50% of the global total comes from erosion associated with high relief on islands of Oceania - a phenomenon which has been underestimated in previous estimates of global sediment production. While erosion on mountainous islands and in upland areas of continental rivers reflects natural topographic influences, Milliman and Syvitski suggest that human influences in Oceania and southern Asia cause disproportionately high sediment loads in these regions.

Sediment, as a physical pollutant, impacts receiving waters in the following principal ways:

- High levels of turbidity limit penetration of sunlight into the water column, thereby limiting or prohibiting growth of algae and rooted aquatic plants. In spawning rivers, gravel beds are blanketed with fine sediment which inhibits or prevents spawning of fish. In either case, the consequence is disruption of the aquatic ecosystem by destruction of habitat. Notwithstanding these undesirable effects, the hypertrophic (nutrient rich) status of many shallow lakes, especially in developing countries, would give rise to immense growth of algae and rooted plants were it not for the limiting effect of light extinction due to high turbidity. In this sense, high turbidity can be "beneficial" in highly eutrophic lakes; nevertheless, many countries recognise that this situation is undesirable for both aesthetic and economic reasons and are seeking means to reduce both turbidity and nutrient levels. The impact of sediment on coral reefs.

- High levels of sedimentation in rivers leads to physical disruption of the hydraulic characteristics of the channel. This can have serious impacts on navigation through reduction in depth of the channel, and can lead to increased flooding because of reductions in capacity of the river channel to efficiently route water through the drainage basin. For example, calculations by the UFRGS (1991) of erosion and sediment transport in the Sao Francisco River Basin, a large drainage system in eastern Brazil, demonstrate that the central portion of the river basin is now dominated by sediment deposition. This has resulted in serious disruption of river transportation, and clogs hydraulic facilities that have been built to provide irrigation water from the main river channel. The sediment largely originates from rapidly eroding sub-basins due to poor agricultural practices.

The most obvious effect of high sediment content on water quality is turbidity. High turbidity can have detrimental effects on phytoplankton productivity because of attenuation of incoming light. If the suspended load has a high organic carbon content, the biochemical oxygen demand will be raised, and conversely, the dissolved oxygen levels will decrease.

Although agriculture contributes to a wide range of water quality problems, anthropogenic erosion and sedimentation is a global issue that tends to be primarily associated with agriculture. While there are no global figures, it is probable that agriculture, in the broadest context, is responsible for much of the global sediment supply to rivers, lakes, estuaries and finally into the world's oceans. Pollution by sediment has two major dimensions.

One is the *Physical Dimension* - top soil loss and land degradation by gullying and sheet erosion and which leads both to excessive levels of turbidity in receiving waters, and to off-site ecological and physical impacts from deposition in river and lake beds.

The other is a *Chemical Dimension* - the silt and clay fraction (<63m m fraction), is a primary carrier of adsorbed chemicals, especially phosphorus, chlorinated pesticides and most metals, which are transported by sediment into the aquatic system.

Erosion is also a net cost to agriculture insofar as loss of top soil represents an economic loss through loss of productive land by erosion of top soil, and a loss of nutrients and organic matter that must be replaced by fertilizer at considerable cost to the farmer in order to maintain soil productivity. The reader is referred to Roose for a detailed analysis of the social, economic and physical consequences of erosion of agricultural land and of measures that should be taken to control erosion under different types of land use, especially in developing countries.

Whereas Roose is mainly concerned with the impact of erosion on agriculture, this publication is primarily concerned with agricultural erosion from the perspective of its impacts on downstream water quality.

Control of agricultural pollution usually begins, therefore, with measures to control erosion and sediment runoff. The principal mechanisms which govern erosion processes, and those measures which can be taken to control erosion.

GLOBAL ESTIMATES OF EROSION AND SEDIMENT TRANSPORT

Sediment as a Physical Pollutant

Global estimates of erosion and sediment transport in major rivers of the world vary widely, reflecting the difficulty in obtaining reliable values for sediment concentration and discharge in many countries, the assumptions that are made by different researchers, and the opposing effects of accelerated erosion due to human activities (deforestation, poor agricultural practices, road construction, etc.) relative to sediment storage by dam construction. Milliman and Syvitski estimate global sediment load to oceans in the mid-20th century at 20 thousand million t/yr, of which about 30 per cent comes from rivers of southern Asia (including the Yangtze and Yellow Rivers of China).

Significantly, they believe that almost 50 per cent of the global total comes from erosion associated with high relief on islands of Oceania - a phenomenon which has been underestimated in previous estimates of global sediment production. While erosion on mountainous islands and in upland areas of continental rivers reflects natural topographic influences, Milliman and Syvitski suggest that human influences in Oceania and southern Asia cause disproportionately high sediment loads in these regions.

Sediment, as a physical pollutant, impacts receiving waters in the following principal ways:

- High levels of turbidity limit penetration of sunlight into the water column, thereby limiting or prohibiting growth of algae and rooted aquatic plants. In spawning rivers, gravel beds are blanketed with fine sediment which inhibits or prevents spawning of fish. In either case, the consequence is disruption of the aquatic ecosystem by destruction of habitat. Notwithstanding these undesirable effects, the hypertrophic (nutrient rich) status of many shallow lakes, especially in developing countries, would give rise to immense growth of algae and rooted plants were it not for the limiting effect of light extinction due to high turbidity. In this sense, high turbidity can be "beneficial" in highly eutrophic lakes; nevertheless, many countries recognise that this situation is undesirable for both aesthetic and economic reasons and are seeking means to reduce both turbidity and nutrient levels.
- High levels of sedimentation in rivers leads to physical disruption of the hydraulic characteristics of the channel. This can have serious impacts on navigation through reduction in depth of the channel, and can lead to increased flooding because of reductions in capacity of the river channel to efficiently route water through the drainage basin. For example, calculations by the UFRGS of erosion and sediment transport in the Sao Francisco River Basin, a large drainage system in eastern Brazil, demonstrate that the central portion of the river basin is now dominated by sediment deposition. This has resulted in serious disruption of river transportation, and clogs hydraulic facilities that have been built to provide irrigation water from the main river channel. The sediment largely originates from rapidly eroding sub-basins due to poor agricultural practices.

Sediment as a chemical pollutant

The role of sediment in chemical pollution is tied both to the particle size of sediment, and to the amount of particulate organic carbon associated with the sediment. The chemically active fraction of sediment is usually cited as that portion which is smaller than 63 m m (silt + clay) fraction. For phosphorus and metals, particle size is of primary importance due to the large surface area of very small particles. Phosphorus and metals tend to be highly attracted to ionic exchange sites that are associated with clay particles and with the iron and manganese coatings that commonly occur on these small particles.

Many of the persistent, bioaccumulating and toxic organic contaminants, especially chlorinated compounds including many pesticides, are strongly associated with sediment and especially with the organic carbon that is transported as part of the sediment load in rivers. Measurement of phosphorus transport in North America and Europe indicate that as much as 90 per cent of the total phosphorus flux in rivers can be in association with suspended sediment.

The affinity for particulate matter by an organic chemical is described by its octanol-water partitioning coefficient (K_{OW}). This partitioning coefficient is well known for most organic chemicals and is the basis for predicting the environmental fate of organic chemicals. Chemicals with low values of K_{OW} are readily soluble, whereas those with high values of K_{OW} are described as "hydrophobic" and tend to be associated with particulates.

Chlorinated compounds such as DDT and other chlorinated pesticides are very hydrophobic and are not, therefore, easily analysed in water samples due to the very low solubility of the chemical. For organic chemicals, the most important component of the sediment load appears to be the particulate organic carbon fraction which is transported as part of the sediment. Scientists have further refined the partitioning coefficient to describe the association with the organic carbon fraction (K_{OC}).

Another important variable is the concentration of sediment, especially the <63 m m fraction, in the water column. Even those chemicals that are highly hydrophobic will be found in trace levels in soluble form. Where the suspended load is very small (say, less than 25 mg/l), the amount of water is so large relative to the amount of sediment that the bulk of the load of the chemical may be in the soluble fraction.

Unlike phosphorus and metals, the transport and fate of sediment-associated organic chemicals is complicated by microbial degradation that occurs during sediment transport in rivers and in deposited sediment. Nevertheless, the role of sediment in the transport and fate of agricultural chemicals, both for nutrients, metals, and pesticides is well known and must be taken into account when monitoring for these chemicals, and when applying models as a means of determining optimal management strategies at the field and watershed level.

For this reason, models using the "fugacity" concept (uses the partitioning characteristics of chemicals as a basis for determining the environmental compartment - air, sediment, water, biota - in which the chemical is primarily found) has proven effective in predicting the environmental pathways and fate of contaminants.

Organic chemicals associated with sediment enter into the food chain in a variety of ways. Sediment is directly ingested by fish however, more commonly, fine sediment (especially the carbon fraction) is the food supply for benthic (bottom dwelling) organisms which, in turn, are the food source for high organisms. Ultimately, toxic compounds bioaccumulate in fish and other top predators. In this way, pesticides that are transported off the land as part of the runoff and erosion process, accumulate in top predators including man.

Key processes: Precipitation and runoff

The major characteristic of non-point source pollution is that the primary transfer mechanisms from land to water are driven by those hydrological processes that lead to runoff of nutrients, sediment, and pesticides. This is important, not only to understand the nature of agricultural pollution, but also because modelling of hydrological processes is the primary mechanism by which agriculturalists estimate and predict agricultural runoff and aquatic impacts. Except where agricultural chemicals are dumped directly into watercourses, almost all other non-point source control techniques in agriculture involve control or modification of runoff processes through various land and animal (manure) management techniques.

In large parts of the world, precipitation is in the form of rain. However, in those areas where precipitation is in the form of snow, the science becomes more complex. Nevertheless, control measures, whether for areas subject to rain or snow can be easily summarized. Therefore, for the purpose of this publication, focus will be on the relationships between rainfall and runoff.

While the practice of hydrology can be quite theoretical, the principal concepts are easily understood.

Rainfall

The primary controlling factor is the rate (intensity) of rainfall. This controls the amount of water available at the ground surface, and is closely related to measures of energy that are used in many mathematical formulations to calculate soil detachment by rain drops. Soil detachment makes soil particles available for sediment runoff.

Soil Permeability

Permeability is a physical characteristic of a soil and is a measure of the ability of the soil to pass water, under saturated conditions, through the natural voids that exist in the soil. Permeability is a function of soil texture, mineral and organic composition, etc.. In contrast, "porosity" is the measure of the amount of void space in a soil; however, permeability refers to the extent to which the porosity is made up of interconnecting voids that allow water to pass through the soil. As an example, styrofoam is highly porous but impermeable, whereas a sponge is both porous and permeable.

Infiltration

Infiltration rate, the rate at which surface water passes into the soil (cm/hr), is one of the most common terms in hydrologic equations for calculating surface runoff. Infiltration is not identical to permeability; it is mainly controlled by capillary forces in the soil which, in turn, reflect the prevailing conditions of soil moisture, soil texture, degree of surface compaction, etc.

Infiltration will vary between and within rainfall events, depending upon factors such as antecedent soil moisture, nature of vegetation, etc. In general, infiltration rate begins at a high value during a precipitation event, and decreases to a small value when the soil has become saturated.

Interflow

Because soil horizons have different levels of permeability not all water in the soil will move downward into the groundwater. The residual water in the soil will move along the soil horizons, parallel to the ground surface. Interflow usually emerges near the bottom of slopes and in valley bottoms. Therefore, identification of these hydrologically active zones is an important part of agricultural non-point source control measures.

Interflow is the mechanism which has also been linked to soil piping, a potentially destructive characteristic in some soils by which shallow "pipes" form naturally in the soil and are enlarged by interflow to the point where they collapse causing gullies in the agricultural surface.

Groundwater

Groundwater is supplied by water which passes through the soil horizons into the parent material and/or bedrock underlying the soil. Groundwater tends to flow towards rivers channels where it emerges and supports stream flow during periods of little or no rain.

This component of stream flow is called "base flow". The chemistry of base flow reflects the soil and bedrock geochemistry, plus any agrochemicals that have been leached into the groundwater.

Surface Runoff

This is the amount of water available at the surface after all losses have been accounted for. Losses include evapotranspiration by plants, water that is stored in surface depressions caused by irregularity in the soil surface, and water that infiltrates into the soil. The interaction between infiltration rate and precipitation rate mainly governs the amount of surface runoff.

Intense rainstorms tend to produce much surface runoff because the rate of precipitation greatly exceeds the infiltration rate. Similarly, in areas of monsoonal rain and tropical storms, the length and intensity of precipitation frequently exceeds infiltration capacity. Destruction of protective surface vegetation and compaction of the soil, especially in tropical environments, leads to major erosional phenomena due to the amount of surface runoff.

Except for nitrogen which is usually found in groundwater in agricultural areas, surface runoff is the primary contributor of agricultural chemicals, animal wastes, and sediment to river channels.

Snowmelt

The phenomenon of snowmelt greatly complicates prediction of agricultural pollution using conventional hydrologic models. Snowmelt, by itself, is not normally a major producer of surface runoff. However, the combination of spring rain and snowmelt on frozen or thawing soils can produce serious erosional problems. Snowmelt tends to contribute greatly to agricultural non-point source pollution by carrying to adjacent streams the animal wastes, sludges, and other wastes that were spread on frozen agricultural soils during the winter period. Correct management of animal wastes in regions of frozen ground has major beneficial effects on water quality.

Sediment delivery ratio

The sediment delivery ratio (SDR) is commonly used in erosion and transport studies to describe the extent to which eroded soil (sediment) is stored within the basin. The SDR is defined as:

$$\text{SDR} = \frac{\text{Measured Sediment Yield}}{\text{Grosserosionin the basin}}$$

where *yield* is determined from reservoir sedimentation or from a sediment monitoring station, and *gross erosion* is estimated using an estimation techniques such as the Universal Soil Loss Equation.

The SDR is always less than 1.0, indicating that soil that is eroded at the field level tends not to travel far before it is deposited. Indeed, sediment storage in rills on fields, at field margins and at the foot of slopes is large. Storage also occurs in river channels (bed and over bank deposition), in wetlands, and in reservoirs and lakes. The SDR is highly variable, however the concept is one of the most important in the understanding of erosion and sedimentation processes and how these operate in time and space.

Sediment enrichment ratio

The concept of the sediment enrichment ratio (SER) is quite important in understanding the impact and economic cost of chemical loss from fields. The process of surface erosion tends to be selective towards fine particles. Consequently, the particle size characteristics of material eroded at source (at the plot level) is progressively changed towards finer particles through deposition of the coarser fraction (e.g. sand-size material).

Because of the chemically enriched nature of fine particles due to the large surface area of clay-size sediment, the concentration of chemicals that are associated with sediment (phosphorus, metals, organic nitrogen, hydrophobic pesticides) increases as the impoverished sand-size fraction is lost during down-field transport resulting in an increasing proportion of the chemically enriched fine (silt-clay) fraction.

The Sediment Enrichment Ratio (SER) is defined as:

$$SER = \frac{\text{Concentration of chemical "X" in transported sediment}}{\text{Concentration of the chemical "X" in the soil}}$$

The importance of the enrichment ratio lies in the fact that there is proportionally more fine-grained sediment transported than coarse-grained sediment during surface erosion. Therefore, the sediment being transported has a finer texture than the source soil material.

Because of the affinity of soil nutrients for fine sediment, this proportionally larger loss of fine material means that there is net impoverishment of the soil. This usually constitutes "mining" of the natural nutrition of the soil (often referred to as "natural capital") and which may never be replaced by the addition of fertilizer. The cost to the farmer is therefore two-fold: loss of productivity due to loss of natural nutrition in the soil; and economic cost of fertilizer which is added in the attempt to compensate for this loss.

WATER QUALITY IN THE DEVELOPING COUNTRIES

Use of pesticides in developing countries is extremely variable, from nil in large parts of Africa, to extremely heavy dosage in intensive agricultural areas of Brazil and plantations of Central America. In their review of the limited research literature on pesticide use and impacts in Africa, Calamari and Naeve conclude that, "The concentrations found in various aquatic compartments, with few exceptions are lower than in other parts of the world, in particular in developed countries which have a longer history of high pesticide consumption and intense use. Generally, the coastal waters, sediments and biota are less contaminated than inland water environmental compartments, with the exception of a few hot spots."

The Brazilian State of Paraná is typical of developing countries undergoing rapid expansion of agriculture, and illustrates the dilemma of pesticide monitoring. Andreoli reports that Brazil, in general, had become the world's third largest user of "agrotoxic" substances by 1970, only exceeded by France and the United States. However, Andreoli states that less than 15 per cent of active ingredients marketed in Brazil are analysed owing to the lack of methodology, equipment and financial resources.

Yet, in a major study of 17 agrotoxic substances (including 11 organochlorine pesticides) over the period 1976 and 1984 in the Paraná River basin, 91.4 per cent of *in situ* samples contained at least one In the Pirapó sub-basin 97.2 per cent of ambient water supply samples and fully 100 per cent of samples from springs showed pesticide residues. Andreoli also notes that studies of "intoxication" in 1985 showed a predominant effect from organophosphate pesticides, with the most serious effects influencing persons between the ages 15 and 25.

The International Code of Conduct on the Distribution and Use of Pesticides, formulated and being implemented by FAO is very relevant to pesticide pollution control and environmental protection in general. The problems of pesticide management in developing countries is somewhat different than those of the developed countries. These are summarized as:

- Inadequate legislation and enforcement of pesticide regulations, including importation, use and disposal.
- Gifts of pesticides from donors that encourage inefficient use and abandonment of older quantities of the same pesticide.
- Stockpiling of pesticides, especially in countries with unstable governments, leading to abandonment of stockpiles in situations of insurrection and civil war. Examples exist where such a situation led to severe groundwater contamination and public health crises due to dumping of pesticides by untrained civilians.
- Storage and handling is a major problem, including leakage from old barrels and deliberate dumping of surplus pesticide mixtures into water courses following application.
- Destruction of old stores of pesticides (due to deterioration of the active ingredient) is financial prohibitive (estimated at US$ 5000 per tonne) especially as stocks must be moved to a developed country for destruction. Consequently, old barrels deteriorate with leakage into surface and groundwater and/or dumping of stocks.
- Lack of training of users in pesticide handling and application, leading to improper application with environmental and public health consequences.
- Use of pesticides for inappropriate purposes, such as killing of trash fish.
- Use of old pesticide drums for drinking water, cooking, etc.

Water pollution is an undesirable change in the state of water, contaminated with harmful substances. It is the second most important environmental issue next to air pollution. Any change in the physical, chemical and biological properties of water that has a harmful effect on living things is water pollution. Water pollution affects all the major water bodies of the world such as lakes, rivers, oceans and groundwater. Polluted water is unfit for drinking and for other consumption processes. It is also not suitable for agricultural and industrial use. The effects of water pollution are harmful to human beings, plants, animals, fish and birds. Polluted water also contains viruses, bacteria, intestinal parasites and other harmful microorganisms, which can cause waterborne diseases such as diarrhea, dysentery, and typhoid. Due to water pollution, the entire ecosystem gets disturbed.

Sources of water pollution

The important sources of water pollution are domestic wastes, industrial effluents and agricultural wastes. Other sources include oil spills, atmospheric deposition, marine dumping, radioactive waste, global warming and eutrophication. Among these, domestic waste (domestic sewage) and industrial waste are the most important sources contributing to water pollution.

Domestic Sewage

Domestic sewage is wastewater generated from the household activities. It contains organic and inorganic materials such as phosphates and nitrates. Organic materials are food and vegetable waste, whereas inorganic materials come from soaps and detergents. Usually people dump the household wastes in the nearby water source, which leads to water pollution. The amount of organic wastes that can be degraded by the water bodies is measured in terms of Biological Oxygen Demand (BOD). BOD is the amount of oxygen needed by microorganisms to decompose the organic waste present in the sewage. The higher the amount of BOD, the more water is polluted with organic waste. Many people are not aware of the fact that soaps and detergents enrich the water bodies with phosphates. These phosphates often lead to algal bloom and eutrophication, which is is most common in stagnant water bodies such as ponds and lakes. Algal bloom and eutrophication lead to the suffocation of fish and other organism in a water body.

Industrial Effluents

Wastewater from the manufacturing and processing industries causes water pollution. The industrial effluents contain organic pollutants and other toxic chemicals. Some of the pollutants from industrial source include lead, mercury, asbestos, nitrates, phosphates, oils, etc. Wastewater from food and chemical processing industries contribute more to water pollution than the other industries such as distilleries, leather processing industries and thermal power plants.

Also dye industries generate wastewater which changes the water quality especially water color. Since the water color is changed, there is alteration in the light penetration and hence it disturbs the aquatic plants and animals. Many of the big industries have come up with wastewater treatment plants.

However, it is not the case with small-scale industries. It is very difficult to treat wastewater from the industries.

Let's take the example of Minamata disease in which more than 1,784 people died and many more suffered due to consumption of fish, bioaccumulated with methyl mercury. It was caused by release of methyl mercury from Chisso Corporation's chemical factory. The disease continued to affect animals and humans for over 30 years, from 1932 to 1968.

Agricultural Waste

Agricultural waste include manure, slurries and runoffs. Most of the agricultural farms use chemical fertilizers and pesticides. The runoffs from these agricultural fields cause water pollution to the nearby water sources such as rivers, streams and lakes. The seepage of fertilizers and pesticides causes groundwater pollution, which is commonly known as leaching. Although the quantity of agricultural waste is low, the effects are highly significant. It causes nutrient and organic pollution to both water and soil. Nutrient pollution causes an increase in the nitrates and phosphates in the water bodies, which leads to eutrophication.

Depending upon the origin, sources of water pollution are classified as point source and non-point source and ground-water pollution. Point source pollution discharges the harmful waste directly into water bodies, for example, disposal through wastewater treatment plants. On the other hand, non-point source pollution delivers indirectly through other ways, for example, water pollution from acid rain.

Prevention of Water Pollution

Although 71 per cent of earth's surface is covered with water bodies, we don't have enough water to drink. Many researches have been done on water purification systems in order to have safe drinking water. However, there are about 1 billion people, who don't have proper excess to drinking water. Therefore, water needs to be conserved and prevent from pollution in order to make it safe for drinking and other consumption process.

Reducing the amount of water use can help conserve water as well as save money. Prevention of water pollution includes using eco-friendly household products such as non-phosphate or low-phosphate detergents and other toiletries, improving housekeeping, turning off the water tap when not needed, disposing the household wastes in proper sites far off from the water sources. Planting more trees can also prevent water pollution by reducing soil erosion and water runoff. Educating people about water pollution is an important way of preventing water pollution.

PESTICIDE MANAGEMENT AND CONTROL

Prediction of water quality impacts of pesticides and related land management practices is an essential element of site-specific control options and for the development of generic approaches for pesticide control. Prediction tools are mainly in the form of models. Also, the key hydrological processes that control infiltration and runoff, and erosion and sediment transport, are controlling factors in the movement of pesticides.

The European Experience

The Netherlands National Institute of Public Health and Environmental Protection concluded that "groundwater is threatened by pesticides in all European states. This is obvious both from the available monitoring data and calculations concerning pesticide load, soil sensitivity and leaching. It has been calculated that on 65 per cent of all agricultural land the EC standard for the sum of pesticides (0.5; m g/l) will be exceeded. In approximately 25 per cent of the area this standard will be exceeded by more than 10 times.

In recognition of pesticide abuse and of environmental and public health impacts the European countries have adopted a variety of measures that include the following:

- Reduction in use of pesticides (by up to 50 per cent in some countries).
- Bans on certain active ingredients.
- Revised pesticide registration criteria.
- Training and licensing of individuals that apply pesticides.
- Reduction of dose and improved scheduling of pesticide application to more effectively meet crop needs and to reduce preventative spraying.
- Testing and approval of spraying apparatus.
- Limitations on aerial spraying.
- Environmental tax on pesticides.
- Promote the use of mechanical and biological alternatives to pesticides.

Elsewhere, as for example Indonesia, reduction in subsidies has reduced the usage of pesticides and has increased the success of integrated pesticide management programmes.

Pesticide registration

Pesticide control is mainly carried out by a system of national registration which limits the manufacture and/or sale of pesticide products to those that have been approved. In developed countries, registration is a formal process whereby pesticides are examined, in particular, for mammalian toxicity (cancers, teratogenic and mutagenic effects, etc.) and for a range of potential environmental effects based on the measured or estimated environmental behaviour of the product based on its physico-chemical properties.

Most developing countries have limited capability to carry out their own tests on pesticides and tend to adopt regulatory criteria from the developed world. As our knowledge of the effects of pesticides in the environment accumulates, it has become apparent that many of the older pesticides have inadequate registration criteria and are being re-evaluated. As a consequence, the environmental effects of many of the older pesticides are now recognized as so serious that they are banned from production or sale in many countries.

A dilemma in many developing countries is that many older pesticides (e.g. DDT) are cheap and effective. Moreover, regulations are often not enforced with the result that many pesticides that are, in fact, banned, are openly sold and used in agricultural practice. The dichotomy between actual pesticide use and official policy on pesticide use is, in many countries, far apart. Regulatory control in many countries is ineffective without a variety of other measures, such as education, incentives, etc. The extent to which these are effective in developed versus developing countries depends very much on:

- The ability of government to effectively regulate and levy taxes
- On the ability or readiness of the farming community to understand and act upon educational programmes.

The fundamental dilemma remains one of accommodating local and short term gain by the farmer (and manufacturer and/or importer) by application of an environmentally dangerous pesticides, with societal good by the act of limiting or banning its use.

There is now such concern over environmental and, in some instances, human health effects of excessive use and abuse of pesticides, that there is active discussion within many governments of the need to include a programme of pesticide reduction as part of a larger strategy of sustainable agriculture. In 1992, Denmark, the Netherlands and Sweden were the first of the 24 member states of the OECD to embark upon such a programme.

The Netherlands is the world's second largest exporter of agricultural produce after the United States. In contrast, wood preservatives in the forest sector account for 70 per cent of Swedish pesticide use with agriculture using only 30 per cent. The lack of baseline data on pesticides in surface waters of OECD countries, is a constraint in establishing baseline values against which performance of the pesticide reduction programme can be measured.

The Danish example: In 1986 the Danish Government initiated an Action Plan for sustainable agriculture which would prevent the use of pesticides for two purposes:

- Safeguard human health - from the risks and adverse effects associated with the use of pesticides, primarily by preventing intake via food and drinking water.

- Protect the environment - both the non-target and beneficial organisms found in the flora and fauna on cultivated land and in aquatic environments.

The objective was to achieve a 50 per cent reduction in the use of agricultural pesticides by 1997 from the average amount of pesticides used during the period 1981-85. This was to be measured by:

- A decline in total sales (by weight) of the active ingredients
- Decrease in frequency of application.

While the World Wide Fund for Nature report that by 1993, sales of active ingredients had been reduced by 30 per cent, the application frequency had not declined.

The Danish legislation included the following components although, by 1993, not all had achieved comparable success.

- Reassessment of active ingredients: Reassessment reflects improved scientific knowledge of pathways, fate and effects of pesticides. By 1993, 80 per cent of the 223 active ingredients had been reassessed. Fewer than 40 per cent had been approved and about 15 per cent are restricted to specific types of application.
- Promotion of organic agriculture: The legislation included funding to promote conversion of traditional agriculture to organic agriculture which, by definition, does not use pesticides.
- Excise tax on pesticides: The Danish Institute of Agriculture concluded that, "A tax on pesticides can be designed and implemented in such a way that it will reduce the use of pesticides without distorting or dramatically worsening the economic situation in the agricultural sector." Funds raised by the tax were to be directed back to the agricultural sector. Studies reported by the Institute of Agriculture suggested, however, that pesticide taxes alone would not produce the requisite reduction during the lifetime of the plan.
- Certification of pesticide users: All farmers and commercial sprayers must hold application certificates. Certification includes education in pesticide issues.
- Approval of spraying equipment: This measure gives the Ministry of Agriculture some control of types of spraying equipment used in Denmark. New computer controlled sprayers permit continuous monitoring of pesticide dose by the farmer and reduces excessive application.

The Danish Government is considering the following additional components as part of the regulatory process:

- *Maximum limits on the environmental load of pesticides:* The intent is to produce an index which equates the quantity used of a pesticide with its known ecological effects. The concept is, however, difficult to implement as noted by the WWF, "... there is no direct relationship between the pesticide-load index and the environmental effects - direct or indirect - of pesticides, since these are the result of a complex interaction between many different factors." Nevertheless, the concept has certain management and regulatory value and may be possible, initially, with a few common pesticides.

- *Prohibiting the use of pesticides within 10 m of lakes, watercourses, wetlands, and conservation areas:* This would achieve some level of pesticide protection for aquatic systems in the same manner that buffer strips are widely used to reduce the effects of sedimentation.
- Prohibiting the use of pesticides within a specified distance from private gardens and properties containing fields that are cultivated without the use of pesticides.
- Prohibiting the use of pesticides within 10 m of a drinking-water reservoir.

The Swedes have had considerable success in achieving pesticide reduction targets. WWF credits their success to the following factors.

- Setting of targets with achievable goals and using multiple measures of reduction.
- Lead role played by the Environment Ministry and Chemicals Inspectorate.
- Active support of farmers organizations which realize the economic and environmental advantages of reduced pesticide usage.
- A strong research and development base that provides credible support for new pesticide initiatives.
- Certification of new machinery and routine testing of farm sprayers at government-regulated test centres.
- Re-evaluation and re-registration of pesticides which has resulted in 338 products being removed from the market.

WATER QUALITY IMPACTS FROM FERTILIZERS

The investigation of eutrophication of surface water by agriculture must adopt a pragmatic management perspective. Of value to agriculture is the perspective adopted by the OECD study of eutrophication. That study focused on the following aspects:

- The qualitative assessment of the trophic state of bodies of water in terms of a few easily measured parameters.
- The dependence of this state on nutritional conditions and nutrient load.

- Translation of these results to the needs of eutrophication control for management.

The progression of these aspects is interesting in that the focus is on easily measurable state of the water body, followed by a determination of the extent to which the state is a product of nutrient loads, then the degree to which loads may be manipulated to achieve a desired trophic state that is determined by water use. Prediction of water quality impacts of fertilizers and related land management practices is an essential element of site-specific control options and for the development of generic approaches for fertilizer control.

Mineral Fertilizers

The response to the need to control leaching and runoff of nutrients and contamination of soils and water by heavy metals has been variable in Europe. Control measures are part of the larger issue of mineral and organic fertilizer usage. FAO/ECE summarized the types of voluntary and mandated controls in Europe that apply to mineral fertilizers as:

- Taxes on fertilizer.
- Requirement for fertilizer plans.
- Preventing the leaching of nutrients after the growing season by increasing the area under autumn/winter green cover, and by sowing crops with elevated nitrogen
- Promoting and subsidizing better application methods, developing new, environmentally sound fertilizers, and promoting soil testing.
- Severely limiting the use of fertilizers in e.g. water extraction areas and nature protection areas.

In any location where intensive agriculture and/or livestock farming produces serious risks of nitrogen pollution, Ignazi recommended the following essential steps that are taken at the farm level:

Rational nitrogen application: To avoid over-fertilization, the rate of nitrogen fertilizer to be applied needs to be calculated on the basis of the "crop nitrogen balance". This takes into account plant needs and amount of N in the soil.

Vegetation cover: As far as possible, keep the soil covered with vegetation. This inhibits build-up of soluble nitrogen by absorbing mineralized nitrogen and preventing leaching during periods of rain.

Manage the period between crops: Organic debris produced by harvesting is easily mineralized into leachable N. Steps to reduce leachable N includes planting of "green manure" crops, and delaying ploughing of straw, roots and leaves into the soil.

Rational irrigation: Poor irrigation has one of the worst impacts on water quality, whereas precision irrigation is one of the least polluting practices as well as reducing net cost of supplied water.

Optimize other cultivation techniques: Highest yields with minimum water quality impacts require optimization of practices such as weed, pest and disease control, liming, balanced mineral fertilizers including trace elements, etc.

Agricultural Planning: Implement erosion control techniques that complement topographic and soil conditions.

Organic Fertilizers

Voluntary and legislated control measures in Europe are intended to have the following benefits:

- *Reduce the leaching of nutrients*
- *Reducing emissions of ammonia*
- *Reducing contamination by heavy metals*

The nature of these measures varies by country; however FAO/ECE have summarized the types of voluntary and mandated control as:

- Maximum numbers of animals per hectare based on amount of manure that can be safely applied per hectare of land.
- Maximum quantities of manure that can be applied on the land is fixed, based on the N and P content of the manure.
- Holdings wishing to keep more than a given number of animals must obtain a license.
- The periods during which it is allowed to apply manure to the land have been limited, and it is obligatory to work it into the ground immediately afterwards.
- Establishment of regulations on minimum capacity for manure storage facilities.
- Establish fertilizer plans.
- Levies (taxes) on surplus manure.
- Areas under autumn/winter green cover were extended, and green fallowing is being promoted.
- Maximum amounts established for spreading of sewage sludge on land based on heavy metal content.
- Change in composition of feed to reduce amount of nutrients and heavy metals.
- Research and implementation of means of reducing ammonia loss.

Sludge Management

Sludge is mentioned here only insofar as the spreading of sludge from municipal wastewater treatment facilities on agricultural land is one method used to get rid of municipal sludge in a way that is perceived to be beneficial.

The alternatives are incineration and land fill. FAO/ECE include sludge within the category of organic fertilizers but note that sludge often contains unacceptable levels of heavy metals.

Economics of control of fertilizer runoff

Nutrient loss is closely associated with rainfall-runoff events. For phosphorus, which tends to be associated with the solid phase (sediment), runoff losses are directly linked to erosion. Therefore the economics of nutrient control tend to be closely tied to the costs of controlling runoff and erosion. Therefore, this will be treated briefly here. In particular, it is useful to examine the economic cost of nutrient runoff which must be replaced by fertilizers if the land is to remain productive.

The link between erosion, increasing fertilizer application, and loss of soil productivity is very direct in many countries. In the Brazilian state of Paraná where agriculture is the base of the state economy, Paraná produces 22 per cent of the national grain production on only 2.4 per cent of the Brazilian territory.

Agricultural expansion in Paraná occurred mainly in the period 1950-1970 and was "characterized by short-term agricultural systems leading to continuous and progressive environmental degradation as a result of economic policies and a totally inappropriate land parcelling and marketing system.

Erosion has led to extensive loss of top soil, large-scale gullying, and silting of ditches and rivers. The use of fertilizers has risen as a consequence, up 575 per cent over the period 1970-1986 and without any gain in crop yields. Loss of N-P-K from an average erosion of 20 t/ha/yr represents an annual economic loss of US$242 million in nutrients.

Analysis by Elwell and Stocking of nutrient loss arising from erosion in Zimbabwe shows similar significant economic losses in African situations. Stocking, applying data collected in the 1960s by Hudson to the soil use map of Zimbabwe, calculated an annual loss of 10 million tonnes of nitrogen and 5 million tonnes of phosphorus annually as a consequence of erosion.

Roose also cites losses of 98 kg/ha/yr of nitrogen, 29 kg/ha/yr of phosphorus, 39 kg/ha/yr of lime and 39 kg/ha/yr of magnesium from soils of lower Côte d'Ivoire as a result of erosion. This loss is so severe that compensation requires 7 tonnes of fresh manure annually, plus 470 kg of ammonium sulphate, 160 kg of superphosphate, 200 kg dolomite and 60 kg of potassium chloride per hectare per year.

Roose notes that it is not surprising that the soil is exhausted after only two years of traditional agriculture. Furthermore, these losses do not take into account additional losses due to harvesting and runoff. Roose summarizes by stating that action against soil erosion is essential in order to manage what he describes as a "terrible" chemical imbalance in soils caused by soil erosion. Estimates of phosphorus loss by erosion in the Republic of South Africa are R26.4 M/yr.

Economic losses tend to be higher in tropical countries where soils, rainfall and agricultural practices are more conducive to erosion, and reported rates of erosion are much above average. The World Bank reported that extrapolation from test-plots of impacts of soil loss on agricultural productivity, indicates some 0.5-1.5 per cent loss of GDP annually for countries such as Costa Rica, Malawi, Mali and Mexico. These losses do not include offsite costs such as reservoir infilling, river sedimentation, damage to irrigation systems, etc.

Soil fertility is a complex issue and nutrient loss is not necessarily nor always a consequence of erosion. Erosion and soil loss is the end member of a variety of physical, vegetative and nutrient factors that lead to soil degradation.

Global patterns of fertilizer application, as reported by Joly, indicate however that rapidly rising levels of fertilizer utilization are required merely to maintain soil productivity from a variety of types of loss, including losses due to erosion and, more generally, to soil degradation.

In a study of 17 agricultural sub-watersheds in the Lake Balaton district of Hungary, Jolankai measured and modelled N and P runoff from a variety of agricultural land uses. He calculated that a selection of control measures (mainly erosion control) would reduce phosphorus loss by 52.8 per cent at a cost of US$ 2500 per ha in remediation measures.

MEASUREMENT AND PREDICTION OF SEDIMENT LOSS

Simple Screening Models

Simple screening models, such as the unit load approach, attempt to provide approximate answers about the likely magnitude of sediment and chemical runoff. This approach is a statistical methodology in which data on runoff of sediment, nutrients and pesticides on a unit area basis (e.g. tonnes of sediment per hectare) are collated from many studies to reflect similarities in crop type, soil and physiographic characteristics.

Unlike the other types of models which are largely focused on prediction and improvement of agricultural management at the farm level, the unit load approach is mainly focused on impacts of agriculture on downstream water quality and without consideration of alternative farm management practices.

Despite the unreliability and large margins of error and related text, this approach has been widely used as a cost-effective means for providing first-approximation answers for agricultural areas for which there are no data. The methodology was originally developed by McElroy *et al.* who collated a major database on unit loads.

This approach was further developed into a US-EPA Screening Procedure by Mills *et al.* and which remains the most comprehensive document on the subject. Unit load data reflect conditions in the United States; application of these data to other climatic and physiographic environments should be avoided. Nevertheless, it is an approach that may be worth considering for development in other parts of the world.

7

Cultural Persistence and Environmental Change

PARALLEL SYSTEMS OF RESOURCE EXPLOITATION

The enthusiastic adoption of sheep-grazing by the Otomí, the profound transformation of the vegetative and soil cover of the Valle del Mezquital under sheep-grazing and the consequent loss of traditional resources, the proximity of the region to the center of Spanish power, and the intensity of contact over several centuries are the same forces that in many other regions transformed "Indians" into "Ladinos." That transformation invites questions with regard to the Otomí. Specifically, how did they remain so distinct? How were they able to adopt a cultural element characteristic of the conquerors-one that was, moreover, a major sector of the evolving colonial political economy-and yet not become ladinos? How were they able to maintain traditional systems of land management when their physical world was transformed literally under their feet?

In reply to these questions, I would propose a mechanism that may have enabled the Otomí of the Valle del Mezquital to adopt pastoralism while at the same time maintaining agricultural practices and approaches to the land that were distinct from Spanish or mestizo systems of land management. This mechanism concerns the formation of ranching as a parallel system of production outside Indian village/ community production. Historically, environmental degradation brought about Otomí marginalization, destroyed the basis of intensive pastoralism itself, and undermined irrigation agriculture using internally generated water.

The combination of political and economic marginalization on the one hand, and physical isolation in a transformed landscape on the other, I suggest, produced a situation that almost guaranteed Otomí ethnic persistence. It should be noted, however, that I am not suggesting that the Otomí remained the same, nor that they did not undergo change; simply, that they remained distinct from Spaniards and mestizos, and, later, Mexicans. I originally constructed the model of parallel systems of resource exploitation as a means to conceptualize the processes by which Spanish pastoralists and their animals were able to expand into densely populated agricultural regions, such as the Valle del Mezquital, without the use of organized force-though with violence, as will be seen. I suggested that the processes by which pastoralism evolved in New Spain were essentially different from those by which Spanish colonial agriculture evolved. Whereas Spanish agriculture was, from the first, embedded in Indian systems of land tenure and use together with village systems of production, pastoralism evolved "outside"-or better, parallel to-village production (Melville 1990).

The initial development of Spanish agriculture occurred within the context of the encomienda system set up immediately following the Spanish defeat of the Triple Alliance, as a system of rewards given to individual conquistadores. As an encomendero, a conquistador was to be supplied with labor and tribute from a specified community of Indians in return for the defense of their customary rights after the collapse of any centralized indigenous authority, within the extremely unsettled conditions of the early conquest era when Spanish authority was insecure. ("Conquest" is here equated with the process by which the colonial regime evolved.)

While the encomienda system clearly served to siphon off tribute to the conquerors, it also functioned as a decentralized system of government. And because it resembled, in broad terms, indigenous tribute systems, it was imposed with remarkably little difficulty.

Although encomenderos had a virtual monopoly on production within the Spanish world up to 1550, primarily by virtue of their access to indigenous labor, they were ultimately dependent on indigenous land-use and land-tenure systems for their agricultural enterprises. They could state the tribute they required and the crops they wanted cultivated, and they could direct their laborers into a variety of entrepreneurial pursuits; but they did not have rights to the lands encompassed by their encomienda communities.9 Crops were grown on Indian lands using Indian labor, very often according to indigenous management systems and schedules. We find wheat being cultivated in the mounds typical of maize agriculture in the early years, for example (Gibson 1964:322).

And later, when it became clear that wheat production on a commercial scale competed with Indian work schedules, most especially maize cultivation, and that the best grains for bread could not be successfully grown during the summer rainy season, the Spaniards changed their scheduling: from spring sowing and fall harvest of rain-fed wheat, to fall sowing and spring harvest of irrigated wheat. As Arnold Bauer (1986) has demonstrated, the shift to a spring harvest meant that threshing was postponed to the following fall, when traditional methods of threshing with the trilla, or with large numbers of mares on a flat area of hardened ground, were not threatened by early rains and did not use animals needed for preparing the ground for maize.

Encomenderos bought or otherwise acquired land (by squatting, grants, gifts, or force) within their encomiendas (Gibson 1964:272-274)-thereby appearing to remove their enterprises from dependence on Indian production systems. But a case can still be made for the dependence of the new agricultural enterprises on the villages by examining developments following the imposition of the repartimiento in 1550. In a move to reduce the power of the encomenderos the Crown took to itself the right to Indian labor and set up government-controlled work gangs: repartimiento in New Spain, mita in the Andes (Gibson 1964:224-236). The shift to the repartimiento system of labor recruitment took the control of Indian labor out of the hands of the encomenderos and made Indian labor available to non-encomenderos, but it did not eliminate the Spanish agriculturalists' dependence on Indian villages; rather, it consolidated this dependence, producing a land-labor system that was carried into the era of the hacienda. Indeed, not only is management of a grain or mixed hacienda inconceivable without Indian villages to supply the work gangs needed to weed and harvest, but the relations between haciendas and Indian villages are treated as a defining characteristic of the hacienda system (Van Young 1983; Gibson 1964:292).

In contrast to the development of Spanish agriculture, the introduction and expansion of pastoralism were not dependent on Indian labor, land tenure, or landmanagement systems. Pastoralists did not need title to a specific area of land, since grass was a "fruit of nature" in Spanish custom, and animals could graze wherever grass grew (Vassberg 1984:6; Chevalier 1975:12).

Nor did they have to rely on Indian labor to care for their animals: they seem often to have relied on black slaves in the early years.10 There was really no reason why anyone who could afford the cost of a few breeding animals and of someone to care for them could not set up a herd, request a grazing license, and be in business. Thus pastoralism evolved, in structural and organizational terms, in parallel to the village-encomienda system of production; it was not embedded within it.

That most early pastoralists were encomenderos, however, is not surprising, since they had the means for capital accumulation and thus the means to acquire breeding stock and develop herds (Miranda 1965; McLeod 1973:46). Nor is it surprising that they often did not bother to obtain licenses, but simply squatted on the lands they needed for corrals and shepherds' huts and grazed their animals in common on the surrounding lands (Gibson

1964:275; Simpson 1952:6; Chevalier 1975: 131). But it is interesting to note that in the Valle del Mezquital, encomenderos grazed their animals in others' encomiendas in the early decades, and rarely in their own: of the fourteen local encomenderos grazing herds of sheep and cattle in the Valle del Mezquital during the 1530s, only three grazed their animals in their own encomienda (Melville 1994b:128-129 n. 25).

Perhaps it was simple pragmatism that kept the encomenderos from grazing in their own encomienda lands. Relations between the herdsmen and the villagers was notoriously violent; indeed, violence seems to have functioned as a means to expand into this densely populated agricultural region and to gain access to forage.

In such circumstances, it would make more sense to annoy someone else's tributaries rather than one's own. But the independence of the encomendero-pastoralists from their own village-encomiendas in the Valle del Mezquital nicely illustrates my point that pastoralism was not structurally and organizationally embedded in the village-encomienda system of production.

As well as serving as a conceptual model for the Spanish invasion of densely populated regions, the model of pastoralism as a parallel system of resource exploitation also serves to clarify the development of the striking differences between the labor regimes of grain and livestock haciendas, and further illustrates the development of pastoralism in New Spain. The marked independence of laborers on livestock haciendas is often ascribed to the low labor needs and less-intensive landmanagement systems of traditional pastoralism (J. Riley 1984:261-262; Zeitlin 1989; Gutiérrez-Brockington 1989:103). I suggest, however, that such differences were not simply the result of the different labor needs of pastoralism and agriculture; rather, they arose in the formative pre-hacienda era as a result of the different developmental processes by which grazing and Spanish agriculture evolved. I base my argument on differences between the labor regimes of agriculture and pastoralism that appeared at an early stage in the Valle del Mezquital. Various categories of wage laborers (sirvientes, "gente," naboríos, gañanes) were resident on the smallscale agricultural and pastoral units of production prior to the appearance of the hacienda in this region.13 But while Indian sheep-shearers living in villages hired themselves out to estancia (sheep station) owners, agricultural specialists did not live in villages and hire themselves out to labor (agricultural units of ca. 200 ha) owners; rather, they lived on the labores in a manner reminiscent of the later hacienda system.

The structural and organizational separation of pastoralism from village production provided Otomí commoners with an opportunity to learn new skills and earn money outside the community-and perhaps to escape increasingly onerous obligations (Zeitlin 1989:47; Hoekstra 1993). It also provided a means for the Otomí elite, and for entire Otomí communities, to engage actively in the new political economy (Melville 1994a). The benefits of raising domestic livestock were many: it was a relatively easy and secure way to obtain needed items such as fat, meat, fiber, traction, manure, and leather. The Otomí evidently preferred sheep over cattle. Several factors shaped their choice of these animals. First, sheep have one great advantage over the other domesticates brought by the Spaniards: their wool; wool is a much warmer fiber than cotton, eminently suitable for the cool highlands, and this would probably have influenced Otomí acceptance of these animals. Second, sheep thrived particularly well in the cool highlands. Third, mutton was preferred over beef in the sixteenth century (Chevalier 1975:142-143). Finally, the development of the Otomí as sheepherders rather than as, say, cattle ranchers was probably shaped by governmental decrees: ganado mayor (cattle and horses) were officially expelled from the central highland regions by viceregal order in the 1550s (except those animals destined for use on agricultural lands), and the region was thus destined to be a sheep-raising region-at least during the sixteenth century (Chevalier 1975:133-135).

The resistance to or acceptance of Old World species was not only based on perceived need, ecosystem compatibility, and government order, however, it also depended on the process by which the new cultural element entered the society. Sheep-grazing was new, it did not replace a traditional system of resource exploitation; and the processes by which pastoralism developed parallel to the villageencomienda system of production meant that it did not unduly disturb the organization of traditional agriculture, even if it helped condense the space for that endeavor. It is an intriguing fact that there was really very little opposition to the initial introduction of livestock, and this despite the fact that cattle and horses, goats and sheep were all grazed within the same space used by the Otomí for agriculture and were corralled next to their houses at night. Livestock raising did not (theoretically) compete with agriculture because grazing animals exploited a previously unexploited (at least from the point of view of the Spanish pastoralists) resource, grass, and pastoralists occupied previously unknown socioeconomic niches.

When opposition to the expansion of grazing did appear, it developed in the context of access to vegetation grown in one particular place: the planted agricultural field. As long as the numbers of domestic livestock remained low, and as long as they ate only grass, there was little difficulty. But domesticated livestock did not distinguish between cultigens and other forage, and Spanish pastoralists were not terribly interested in maintaining indigenous agriculture at the expense-as they saw it-of their rapidly increasing herds, unless forced to do so by viceregal order.15 Furthermore, the fluid power relations inherent in the conquest situation meant that, wherever possible, the Spaniards ignored both Indian needs and viceregal orders and let their animals forage in planted fields.

The Otomí agriculturalists, at least at first, had no knowledge of their rights vis-à-vis the pastoralists, and could not effectively defend themselves against the invading animals. Viceregal orders, such as those of Viceroy Velasco in the 1550s, attempted to codify the timing and use of agricultural fields by grazing animals (Chevalier 1975:133-135). But it was really only with time and experience and many lawsuits-and, most importantly, with the involvement of indigenous peoples such as the Otomí in livestock raising-that the relations between agriculture and pastoralism were codified by local custom and use.

Far more threatening to Otomí agriculture than the actual adoption of pastoralism was the process of environmental transformation that accompanied the expansion of sheep into the Valle del Mezquital.

Forest clearance, burning, and overgrazing degraded the catchment area of the Valle del Mezquital and destroyed the basis of initial Otomí success in this arid region. As the water table dropped and springs failed, landowners and villagers fought each other in and out of court over the reduced water sources, drawing up agreements that usually spelled out the weakening power of villages to control this essential resource. The extent to which the catchment value had deteriorated in the Valle del Mezquital by the end of the sixteenth century is indicated by the fact that the Otomí had to fight for water to irrigate fields that were vastly reduced in extent as a result of the demographic collapse (ca. 92 percent by 1600).18 But, to the considerable chagrin of the Spaniards, the Otomí, though reduced in number and with reduced agricultural fields, managed to retain control over the best lands in the region: the humid bottomlands.

The environmental degradation that accompanied the development of intensive sheep-grazing in the Valle del Mezquital resulted in the deterioration of the carrying capacity of the range and the collapse of intensive pastoralism. Small-scale intensive sheep-grazing, with large numbers of small landowners grazing huge numbers of animals in common, developed in the 1550s and 1560s, reached its apogee in the 1570s, and began to decline rapidly thereafter; by the last two decades of the sixteenth century, much more land was needed to maintain a single head of sheep. The only way to maintain profitable flocks was to acquire several stations and to enforce exclusive access to the pasture. Those who could afford to buy land did so, and by 1600 regional production was dominated by extensively grazed latifundia. Pastoralists who had been known as señores de ganado (livestock owners on a grand scale) in the period of expansion, were now known as terrateñientes (landowners).

Another consequence of environmental transformation was a shift in regional production from sheep to cattle and goats in the seventeenth century (Melville 1994b). But the Otomí did not take part in either cattle-raising or goat-herding on the same scale as they had sheep-grazing in the era of expansion. Indeed, their marginalization can be ascribed, in part, to the depression of sheep-grazing in this region after the sixteenth century. Restrictions on the number of animals Indians could graze, and on the number their abbatoirs could butcher, also limited the ability of the Otomí to actively engage in the pastoral sector of the economy by the end of the century (Dusenberry 1963:111; Chevalier 1975:138).

In the early stages of contact with the Spanish, the Otomí had taken advantage of opportunities provided by a rapidly changing political economy in order to maintain their economic and political status in relation to the newcomers, and perhaps to leave old obligations behind. The evolution of pastoralism as a parallel system of resource exploitation allowed them to adopt livestock-raising while maintaining traditional agricultural systems. The importance of this is very clear when one considers the problems that the Spaniards faced when they tried to introduce wheat production. The Indians of New Spain persistently and successfully resisted the efforts of the Spaniards to force them to grow wheat. It was even difficult to have wheat grown as an item of tribute.

Although wheat was as crucial an element of Spanish culture as sheep- or hog-raising, and was as profitable a sector of the colonial economy, it was not readily accepted. Wheat was not an add-on in the way livestock was-it was a replacement. As a replacement it required extensive changes in many areas of social and cultural life, so that in most places it was rejected (Gibson 1964:322-323).20 But livestock could not simply be added to the culture and landscape of the Valle del Mezquital: pastoralism brought with it profound changes in the environment that, in the end, destroyed the basis of small-scale intensive pastoralism and, as well, the basis for Otomí participation in the colonial political economy, and produced a landscape within which the Otomí villages became islands in a sea of mesquite. Finally, then, one may argue that the persistence of Otomí land-management techniques after the sixteenth century resulted from cultural involution and structural-and physical-isolation.

ENVIRONMENTAL FACTORS IN THE CHRONOLOGY OF HUMAN EVOLUTION AND DISPERSAL

This is a four-year mini-thematic programme aiming to integrate and strengthen UK research in human evolution. The rationale underlying the proposed research and the timing of this programme is the acute awareness that studies of human evolution need to be tied closely to the ecological and environmental expertise underpinning Quaternary science generally, and that in particular research among the disciplines of palaeoanthropology, archaeology and palaeoecology needs to be integrated in an effective manner to realise these wider aims. This programme will be highly interdisciplinary and has two parallel pathways. Firstly it will acquire and analyse new field data, thereby making a major contribution to human evolution and quaternary science. Secondly it will exploit existing datasets from archaeology, palaeoanthropology, palaeoecology, as well as genetics, within a developing E-Science context.

The pathways lead to several key issues relating to a fundamental, curiosity driven question: " What impact did environmental factors have on human evolution and dispersal?

EFCHED aims to explore the interface between the disciplines engaged in human evolution research. These include the classic ones of human evolutionary studies-palaeoanthropology, archaeology and palaeoecology-as well as emerging ones such as evolutionary genetics and palaeoclimatic modelling, as well as allied disciplines such as neoecology (Figure 1). The relationship between environment and evolutionary change will be at the heart of EFCHED projects. Proposals will be judged by their commitment to the role of interdisciplinary approaches in investigating such fundamental relationships.

The mini-thematic programme will serve as a catalyst for a new integrative science within the field of human evolutionary research, and will provide training opportunities for the interdisciplinary postgraduate and postdoctoral researchers that the field urgently requires. The programme's output will be of wide public interest and serve as a showcase for NERC science policy and process.

BACKGROUND

The background for this programme is:

- The UK has an internationally-recognised group of researchers working in this field. Our expertise is particularly strong in human morphology and palaeontology, Palaeolithic archaeology, Quaternary faunas, science-based dating, ancient biomolecules and human genomics. In all these fields we have led the way in developing the environmentally-based approaches that are now becoming a priority.
- The growth in evidence related to both human evolution and its environmental context leading to a need for more large-scale projects that can pull together case studies.
- The growth of molecular approaches to evolution which have opened

up new issues (such as the debate about hominid and human origins) which depend upon timing and environmental context for resolution.
- The development of major new techniques, especially relating to environmental context and reconstruction, that have greatly expanded the data available and approaches possible to human evolution.
- Recognition that the relationship between humans and their environment needs consideration of organismenvironment interactions at both ecological and evolutionary time scales.
- The great public interest in human evolution, evidenced in major TV documentaries, books and newspaper articles.
- The opportunity to exploit the growing availability of major palaeoanthropological and archaeological databases, programmes such as Rapid Climate Change (RAPID), and to bring a greater analytical dimension to studies of human evolution.
- The growing interactions between archaeologists, palaeoanthropologists, palaeoclimatologists and molecular biologists and the need to provide a new disciplinary framework in which they can work together.

NEED FOR A MINI-THEMATIC PROGRAMME

The EFCHED mini-thematic programme has a number of advantages over non-thematic funding:

- Providing a focused scientific response to the challenges posed by recent developments in human evolutionary studies, all of which have a strong environmental component.
- Providing the framework for strengthening long-term research on human evolution.
- Integrating the three essential components of human evolutionary research-primary field projects generating original evidence, laboratory analysis, and synthesis.
- Providing a sharper focus on evaluating and addressing key evolutionary questions.
- Providing a formal link between two fields that have grown together over the years-i.e. human-centred evolutionary and archaeological studies and quaternary science.
- Allowing greater exploitation of emerging scientific techniques, especially science-based dating, ancient biomolecules, evolutionary genetics, and palaeoenvironmental/climatic reconstruction.
- Providing a better way of integrating the development of scientific techniques (one of the great strengths of SBA) within problem-driven research.

Recent Achievements in Human Evolutionary Studies

Listed below are 10 examples of key discoveries and developments in human evolutionary studies that prompt environmentally-based approaches:

1. The recognition of more than 15 species in the hominid fossil record, leading to questions about the environmental and climatic factors leading to speciation, extinction and isolation.
2. The fact that bipedalism and brain enlargement evolved independently, at widely separated times, and in assorted environmental conditions.
3. The identification of a major shift in body size and shape with the evolution of Homo, compared to Australopithecus, with the possibility that this is related to a major shift in the way the environment was used.
4. The discovery through isotope analysis that the australopithecine diet may have been far more varied than had been inferred from classic morphological studies.
5. The realisation that humans first occupied almost three quarters of the earth during only the last one per cent of the time since hominids first appeared
6. The recognition that stone tools remain remarkably stable over long time-spans, and that technological solutions to survival among earlier hominids were not habitat specific. This implies that human culture is contingent upon both environment and phylogeny.
7. The finding that there was a significant level of geographical isolation and evolutionary divergence among Middle Pleistocene hominids, and that this may have been related to changing environmental patterns associated with the onset of glacial cycles.

   EFCHED: proposal for a NERC thematic programme 07/02/2003 3
8. The recognition, derived largely from molecular genetics that hominid evolution may have been characterised by extreme demographic instability, and that climatically driven population contractions and expansions may have given rise to our own species.
9. The proposal that the early divergence of modern human populations may have been influenced by volcanic as well as climatic events.
10. The discovery, through both archaeology, fossils and genetics, that the origins of modern humans is a localised African event, closely connected to the last two glacial cycles.

RESEARCH AIMS AND METHODS

The aim of the proposed theme is to encourage research that formulates and tests hypotheses on several chronological and spatial scales that bear on the causal relations between environmental and climatic change and human evolutionary patterns-genetic, morphological, behavioural and cultural. These relationships are epitomised in the questions posed at the outset, and from which more specific hypotheses can be derived-i.e. 'what impact did environmental factors have on human evolution and dispersal?'.

The hypotheses will be tested through interdisciplinary projects that will include:

- The compilation of relevant archives and further enhancement via fieldwork, collections analysis and experiment.
- The analysis of evidence at a variety of geographical and temporal scales using a number of approaches including modelling, and comparative and experimental studies.
- New fieldwork to generate new data to answer specific questions
- Development of new syntheses based on comparison across time scales and regions.

RESEARCH THEMES

We identify two major related themes as lying at the heart of the EFCHED programme: " Humans in an evolutionary context " Humans and dispersals. Each of these sub-themes poses research problems that involve archaeologists, palaeoanthropologists, palaeoecologists, palaeontologists, molecular geneticists, as well as earth and life scientists more generally, although for each of them the mix and emphasis varies. These themes drive the selection of areas and time periods which will deliver appropriate data sets to achieve the programme aim.

Humans in an Evolutionary Context

It is second nature to archaeologists and palaeoanthropologists to be aware of the long term and dynamic nature of the human lineage, but this is still something that has not been operationalised through into many dis ciplines.

Looking at humans in an evolutionary context-i.e. as a series of different species, with different morphologies, behaviours and ecological contexts-and using the excellent archaeological and fossil record documenting hominid presence (and absence) over a period of nearly 5 Myr allows a number of issues to be explored:

- Macro and microevolution: some hominid species living more than 4 million years ago were very ape-like, and their record is no different from that of any other mammalian Cenozoic lineage; others, such as Neanderthals, are richly recorded in both tools and bones, and have a span of only 200,000 or less. The process of transformation from one sort of creature to the other (when and how did it occur?) will be a challenge to many theoreticians and empirical scientists.

- The wealth of the hominid fossil record permits issues such as the timings of speciation and extinctions to be examined.
- The fossil record records the history of human morphological adaptations, and this can be used both to examine such major evolutionary issues as rates of change and also the actual nature of human biological adaptation.
- The archaeological record (e.g. stone tools) provides the potential to study lineage persistence and distribution, and behavioural adaptation, not found in other species. As a result it potentially provides one of the best test-beds for general evolutionary hypotheses.
- A critically important and rapidly emerging field is that of evolutionary molecular genetics, and this is being used to reconstruct the phylogeny of humans and other species, to estimate coalescence events, and reconstruct prehistoric demographic patterns. However, there is considerable controversy about the underlying chronological assumptions, based on mutation rates and other internal inferences. The Pliocene and Pleistocene fossil records can be used to provide a broad comparative dataset through which calibration of molecular systems can be tackled.
- The development of bioinformatics has led to the potential to develop more sophisticated and analytical uses of chronological data (e.g. radiocarbon lists, etc.). This database approach ('palaeoinformatics') can be used to examine in detail the chronological patterns of human distributions in the environment.
- The period in question (5-0 Myr) covers times in which very different chronological resolution is possible, from the early part where chronology is measured in 105y to the recent end, where a resolution of 102y may be possible. Within these scales is the fact that some archaeological context s allow 15 minute activity episodes to be observed. One of the tasks for an evolutionary explanation lies in the integration of different time scales.
- The growth of palaeoinformatics allows testing across dating methods and scales, and thus develops a more stable chronological framework for scientists from a number of disciplines.
- The impact of rapid climatic change on human society, as revealed for example in the chronological precision now provided by the Greenland ice cores, can now be investigated at a variety of scales and the outcomes observed.

Humans and Dispersals

The course of human evolution sees hominids changing from a group of geographically restricted African species to a single global species. There is thus a strong geographical, and therefore ecological, element to hominid evolution, and a key aim will be the investigation of the patterns and processes by which hominids repeatedly dispersed and contracted across landscapes at all scales.

A major question to ask is, what role did environmental change play in the observed pattern? Understanding the geological context at a variety of scales and processes provides the critical framework for this theme

- Dispersals are strongly influenced by environmental effects, and it will be a major part of the theme to investigate the timings and distributions of dispersals in relation to local and global patterns of climatic/ environmental change.
- Dispersals are also strongly influenced by the behaviour of the hominids, including their technology, and comparing patterns of dispersals across different human populations will show the role of changing behaviour in the distributions and persistence of human groups.
- Dispersals and contractions form the basis for biogeographical patterns, for both humans and other species. Allopatry, isolation, and admixture are all major evolutionary factors, and identifying corridors of dispersals, refugia, and contact zones will be of use to ecologists, geneticists and other scientists.
- As populations disperse-or contract into refugia-they come under new selective influences and contexts for genetic drift. Hominid morphology thus has a strong geographical component that can only be understood by bringing together ecological factors with the historical contingencies of the populations.
- At the macro-level it has been proposed that pulses of speciation and extinction have been directly related to major climatic change. The theme will explore this issue in terms of geographical variation, taxonomic differences, and rates of change.
- Embedded within global patterns of change are local processes, varying from regional climatic patterns, shifts in sea and lake levels, and tectonic/ volcanic effects. These are the ones that directly impinge on human and other populations, and one focus will be the discerning of such local patterns within larger scale changes.
- During most of human evolution the direction of influence has been the environment acting as the selective pressure shaping human morphology and behaviour. The morphological and behavioural

patterns observed in the fossil record will be screened for the effects of environmental change.

- With the emergence of modern humans and the global expansion of population the direction of causality shifted, and humans increasingly influenced their environment. Mapping this process and its effects within a geological framework will be a major area of research

RELEVANCE TO NERC: NATIONAL AND INTERNATIONAL DIMENSIONS

NERC's Strategy for Science-Based Archaeology identifies three key initiatives which are directly addressed in this programme; Refining the chronology of the human past, Human evolution, ecology and behaviour and biomolecular archaeology. The proposal fits squarely into two of the six "science opportunities" identified by the Science and Technology Board sectoral strategies which are: Climatic and Environmental Change and Evolution of the Biosphere and the global environmental change area of core science. The programme builds on the previous SBASG Thematic Programme , the Ancient Biomolecules Initiative (ABI) and dovetails with the new RAPID programme. Recent NERC initiatives in E-Science will also impact on this programme.

Nationally, the Prehistoric Society in its 1999 report "Research Frameworks for the Palaeolithic and Mesolithic of Britain and Ireland", sponsored by English Heritage, identified the issues of dispersal as being central to future research. This report has been acted upon by English Heritage, which has commissioned further scoping studies for research investment. The Stage 3 Project, run through the Godwin Centre at Cambridge, provides a successful model for international, interdisciplinary research.

The international importance of human evolution is demonstrated by the National Science Foundation (USA), "Human origins initiative" currently under consideration as a ten year proposal. European interest in human evolution research is delivered, among others, through the ESF, CNRS (France) and the Max Planck Institute (Germany). UNESCO's World Heritage List contains sites relevant to human evolution in four continents. INQUA now has a sub-commission on climatic imperatives and historical contingency.

UTILITY AND BENEFICIARIES

A prime benefit of this Programme would be an increased and intensified interaction between diverse scientific disciplines encompassing a wide range of environmental, biological and historical sciences. Of necessity, this requires a world-wide perspective and offers exciting possibilities for international collaboration. It also demands explicit recognition that the collaboration of experts in fields outside archaeology and anthropology is more effective when the various investigations are formulated to include objectives that are of direct interest to the supporting sciences. The co-ordination through EFCHED of these potential co-funding partners will benefit NERC by extending the application of skills and the shared use of facilities and equipment. The proposal builds on recent NERC, JIF and Wellcome investment (the Radiocarbon Accelerator and Ancient Biomolecules Centre at Oxford) the AHRB Research Centre for the Evolutionary study of Material Culture at UCL and Southampton and Cambridge's £5m award from Wellcome, Leverhulme and SRIF to establish a Centre for Human Evolutionary Studies, all of which offer the potential for training PDRAs with an interdisciplinary focus.

Furthermore, the proposed programme will be of direct benefit to all scientists interested in human response to global climate change, and ultimately to policy makers. It will also be of direct relevance to organisations such as English Heritage, currently funding internationally important investigations of early hominid occupation on the Sussex coastal plain, and who are charged with managing and presenting the human heritage. EFCHED funded projects will be expected to produce materials which bring human evolution alive for the public. Involvement with the media to explain the process of science and the role of research councils such as NERC is seen as important to the programme.

TRAINING

The opportunity to train Postgraduate and Postdoctoral researchers within an interdisciplinary research programme is a key objective of EFCHED. The programme will break down disciplinary boundaries both through the requirements to train skilled practitioners in the structure of the varied data archives, and also in the range of scientific techniques and methods which are needed to interrogate them. The programme provides the potential to shape the next generation of researchers and equip them with the skills and interdisciplinary dialogues that only come through engagement with cutting-edge science. This will be achieved by tied studentships and project based PDRAs. It is to this cohort that we look for the new disciplinary framework focused on integrated human evolution rather than its current constituent parts. We predict that such a lead will also initiate a sea change in the structure of graduate training programmes. In particular we envisage inter-institutional collaboration that will share expertise and broaden the student experience. Human evolution is an international subject and we see opportunity through such initiatives as the World-Wide University Network, linking UK and US universities, to further demonstrate this claim.

ART: FUNCTIONAL BEAUTY AND COMMERCIAL COMMODITY

Art in the Euro-american understanding of the term-objects made solely for aesthetic purposes-did not exist in the Arctic until recently. However, an archaeological record with wonderful sculptures and drawings shows that peoples in the Arctic have been making objects that were functional and aesthetically pleasing from time immemorial.

The first encounters with outsiders provided new possibilities for artistic expression, for example by the introduction of iron tools. However, Christian missionaries and government officials were often responsible for undermining the religious basis on which most of 58 Arctic Human Development Report The Nenets sacred site Numt' khantorma.

Subsistence, narrowly defined, means to survive. To the Gwich'in, it means far more. Besides our spiritual relationship with God, the Creator of all that is, subsistence is the essence of the Gwich'in Nation. It is how we are sustained physically. It serves to support us economically and spiritually and is a key to our sustainability as a people. We are fed by plants and animals of the water, air, and land. Wood provides warmth and housing and the raw materials for tools and transportation devices, such as boats, snowshoes, and toboggans [runnerless sleds widely used by Native Americans].

The cash component of our economy is a small but important element of our subsistence lifestyle. We harvest wood, fur, and wildlife for barter or sale, mostly locally, but occasionally abroad. While some have assumed the role of full-time workers, many rely mostly on natural resources they catch or cut to meet their economic needs. Our spirituality is tied inextricably to the water, land, air, and resources within them. Our relationship to the spiritual realm has always been conducted through the beauty and awe of nature around us. We are sustained as individuals and as a people through our subsistence lifestyle. Our resourcedependent culture is sustained by unfettered mobility and access to the resources we depend on. Unfettered mobility means that we pursue resources where they are most abundant and where there is easy access. Unfettered access means that we gather resources when we need them and in the most economical and feasible manner possible. As we continue this lifestyle, we build and maintain our close and interdependent ties to the resources around us.

Gwich'in identity is a picture of integration with the land and resources. We see ourselves as an integral part of the diversity of the landscape. We believe that we would not be whole if we were separated from this land. We also believe that this land would not be whole without our presence. Our well-being is linked closely with our ability to live on and adapt with the land. Our family and land-based bonds are strengthened, restored, and invigorated as we continue our subsistence lifestyle. A tremendous sense of belonging and purpose is experienced as we survive on the land. Subsistence encompasses all areas of Gwich'in life from the cradle to the grave. Gwich'in youth are trained early on the intricacies of our relationship with the natural environment and the harvesting, processing, and distribution of wild resources. These relationships are strengthened when our youth mature into leaders. The cycle of life continues as they pass what they have learned from their elders and through trial and error to their children and relatives.

Arctic artistic production was based. In certain areas, such as Greenland, art came under direct influence of European traditions early on. The Greenlander Aron of Kangeq (1822-1869) became known throughout Greenland and Denmark for his lively watercolors of Inuit village life and tales. In other areas, such as Alaska and many parts of Arctic Canada, handicraft items for trade provided a venue for "native art." For example, delicate Athabascan beadwork on moose and caribou skin was popular throughout the Canadian and Alaskan Arctic and subarctic.

The entry of Arctic art into international markets is recent. One of the best-known examples is Canadian Inuit printmaking. In 1948, James Houston, a young non-Inuit Canadian artist, traveled north to the Nunavik village of Inukjuak for a sketching trip. Houston befriended the local Inuit, who coveted imported commodities. In trade, the Inuit brought him small soapstone models of animals.

Houston persuaded the Canadian Government to subsidize soapstone carving, which eventually became a multi-million-dollar enterprise for the Inuit. A decade later, Houston had moved to Cape Dorset on Baffin Island and repeated the same success story with printmaking. There, local Inuit artists submitted drawings for print making. The prints were marketed in North America and Europe, and the demand soon outstripped the supply. Thanks to worldwide media coverage, artists such as Kenojuak Ashevak and Pudlo Pudlat became famous with Inuit art collectors. Their works are in museums, art galleries, and private collections around the globe.

In the early 21st century, indigenous art in the circumpolar North is thriving. Cruise-ship passengers and other tourists are eager to bring home objects which signify the exotic Arctic. In Alaska and coastal British Columbia, gift shops routinely sell copies of native art mass produced in Asia where labor is cheap. The authenticity of indigenous art is to some extent protected by subsidized programs that provide artists with a sticker guaranteeing the authenticity of their work.

However, more and more artists in the Arctic do not want to be seen as representatives of a particular ethnic tradition but as active participants in a globalized art scene. Whatever the position of the individual artist is, the fact remains that almost all indigenous art from the Arctic is today created for consumption in a culture that is economically and politically more powerful than theirs.

In recent years, the development of Arctic arts has gone far beyond the confines of what have been traditionally considered the fine arts. New art forms, such as literature and filmmaking, have become prominent. For example, the critically acclaimed film Atanarjuat ("The Fast Runner")-written by Paul Apak Angilirq and directed by Zacharias Kunuk-is the first feature film made in Inuktitut. Moreover, writers such as the Chukchi novelist Yuriy Rytkheu have successfully transformed oral traditions into books which are read throughout the Arctic and non-Arctic world. Finally, new forms of Arctic music are developing, which incorporate traditional elements, such as the Sami yoik, and elements of western popular music.

Similarities and Differences with Non-Arctic Areas

Many of the cultural trends in the Arctic are the result of an unbalanced encounter between the cultural traditions of small-scale, hunter-gatherer societies and large-scale agricultural and industrial states. What is peculiar for the Arctic is that these encounters occurred relatively late, and that agricultural/industrial cultural values were imposed in the 20th century. The similarities to non-Arctic areas are greatest with those of other hunter-gatherers pushed aside by agriculturalists relatively recently, as in Australia and Amazonia.

However, the indigenous groups in the Arctic are generally less impoverished than in their thirdworld counterparts. And even more important, they are part of larger societies that have come to support-by and large-a fuller implementation of civil and indigenous rights.

Variations within the Arctic

Various parts of the Arctic came into intense contact with cultural agents from the outside at different points in time, which in turn often determines the extent to which non-Arctic elements have been incorporated into local cultural traditions. An example is the almost complete erasure of shamanistic elements from Saami worldviews as a result of almost 1,000 years of Christian influence. For current cultural processes in the North, government policies are among the most important variables. In the 20th century, the policies implemented by the Soviet Union differed most from other Arctic countries.

Moreover, the cultural trajectories of Iceland and the Faroe Islands are noticeably different from the rest of the Arctic, primarily because of their different settlement history. While the cultural background of the ancestors of the contemporary Icelanders and Faroese was undoubtedly non-Arctic and agricultural, their descendants can point to over 1,000 years of cultural development in the Arctic.

Trend Summary

Outsiders and Arctic residents have been bemoaning "culture loss" for decades. This kind of judgment fits with the measurable decline in linguistic and religious knowledge, the fact that certain songs, dances and other art forms were pushed out of use, that languages became extinct, and worldviews replaced. However, "culture gain" and "culture creation" are also part of the cultural realities of the Arctic. Vocabularies, dialects, and languages were replaced by others, as were religions and art forms. Also, many aspects of Arctic worldviews have persisted despite processes of change and replacement. In the final analysis, the most important factor is whether the local community in question identifies with the cultural bricolage its residents hold today. Culture is intimately tied to identity and the major question is whether you can consider the language/s you speak and the spiritual entities you respect as "yours," no matter where they "originated."

Socialization, Kinship, and New Networks

The accelerated change of recent decades has been accompanied by gaps in socialization and knowledge transmission between generations. Despite these difficulties, kinship and family relations, especially for indigenous peoples, have remained a central focus. This is true also in the growing Arctic towns and even in urban settings outside the Arctic homeland. Contemporary social networks also include connections between recent Arctic emigrants to more southerly cities and the Arctic communities they came from, as well as between southern immigrants to the Arctic and their social milieu of origin. This section highlights how current social change is shaped by relations between generations, kinship networks, urbanization, and the extension of local communities through emigration.

SOCIAL TRANSMISSION OF KNOWLEDGE ACROSS GENERATIONS

Among indigenous peoples, recent changes have been so precipitous that they have been interpreted by outsiders as a break from the past, and in some cases even, as a breakdown of societies and cultures. In particular, transformations have been associated with major communication gaps between generations. These gaps were amplified by just as abrupt linguistic shifts, which are described in the previous section of this chapter. In many regions, children were in boarding schools where the only language they spoke was a language different from that of their people. Communication with their grandparents was thus often precluded. Arctic developments confirm the fact that minority languages can be dealt deadly blows in the space of one or two generations.

The break in socialization was accentuated when social changes and the school system disrupted important features of traditional education systems. For example, among many indigenous peoples, a great deal of learning occurs through observation and imitation rather than through the written or spoken word. Transmission of knowledge also takes place through the telling of myths and stories that implicitly teach lessons of life to those who are culturally trained to decode them. Furthermore, among Inuit for example, children are considered as autonomous beings, embodying the name soul of a deceased relative, whose own volition should not be interfered with.

These aspects of traditional culture have clashed with the recently introduced formal education and new role models. The introduction of culturally sensitive curricula should reconcile public education with other modes of socialization. In some cases, parents thought they were improving their children's chances by letting them be educated in another culture and language, rather than encumbering them with a cultural heritage they were told was obsolete. Such a break in communication between generations, which was common in the second half of the 20th century and continues today in some regions, has created a bottleneck for social reproduction. Nowadays, parents struggle with the atomization of the household and the competition of television programs when they try to provide role models, while indigenous elders strive to reach out to youngsters. The box below illustrates this modern dilemma.

The Extension of Arctic Rural Communities

Urbanization is a universal and well-studied trend. In the Arctic, where living off the land dictated low population densities and scattered settlements, it is a relatively recent phenomenon, associated with the rise of welfare state policies, industrialization, and the spread of wage employment.

The concentration of the population in permanent settlements took place during the twentieth century. Many of the major Arctic towns have been growing at a rapid pace, and they are becoming increasingly multi-cultural . The figure on the following page provides an illustration of the rapidity of urbanization in Greenland since 1950.

In the past few decades, increasing numbers of indigenous people are also settling down in larger centers away from their home areas. For instance, Oslo, Stockholm and Helsinki are playfully referred to as the largest Saami villages or siidas in the Nordic countries . According to a recent study , about 7,000 Greenlanders live in Denmark, which is equivalent to about 15% of the

Greenlanders in Greenland. Two thirds of them are women, and they are spread throughout the country rather than concentrated in the capital. In 2001, about 10% of Canadian Inuit lived outside the Arctic . Among the recently better-studied Native urban communities are the Yup'ik and Iñupiat in Anchorage, Alaska . In 2003, about 10% of the 274,000 inhabitants of Anchorage were Native or part Native people, which corresponds to almost 17% of the total Native population of the State .

According to Fienup-Riordan, outmigration to cities does not mean the severance of ties with home communities: "Yup'ik communities are not disintegrating, their lifeblood gradually seeping away. Many can be seen as actually In Greenland, family is important also for many people living in towns.

For example, in Nuuk where family members have to work many hours during the week, it becomes important for kinsmen to spend time together in their spare time, during holidays, family celebrations, and times of crisis. Also, because people live in a large-scale society, they have a need to stick together to handle outside crises. Voices gathered in research on the role of kinship among families in Upernavik, with 1,218 inhabitants, and Nuuk with 13,884 inhabitants illustrate this well: "A family is a person whom you can get support from, I don't know what I should do without my family. I need my family-it can be my own family or my husband's family." (35-year old woman).

"Generally most of the families keep together, but a lot of the time they have to work. Your kinsmen give you strength and help you through crises, it is therefore important that families keep together. If families don't stick together it will be harder for them to handle crises, because they will have no one to support them." (67-year old woman)

"I think families are more divided today maybe because they are so busy, but there are still families that are strongly connected. It is very important that families are connected [...] we teach our children to understand how important it is for us to keep together." (young mother, 35 years old) Although urban kinship can be utilized in many different ways, the basic ideology is centered around expectations about moral and mutual obligations in the realms of sharing, naming, and adoption, among others.

Relatives have many different kinds of mutual obligations in their lives and sharing defines the family. Family members strive to keep relatives together to avoid being isolated and having no one to support and share with. People have a choice as to whether or not to fulfill the mutual obligations of the family, and sharing does not always mean that kinship systems and forms of social organization are harmonic.

Gitte Tróndheim, University of Greenland, Nuuk Kinship in urban Greenland expanding and recreating themselves in unprecedented ways until today, when they are as strong and vital as at any time in their 2,500-year history. Although geographically much farther-reaching than ever before, contemporary Yup'ik community continues to be characterized by wide sharing throughout the extended family, with money from wage labor and commercial fishing used to support a variety of harvesting activities to fulfill extended family needs" .

Also, Yup'ik identities are not abandoned, but reinforced in Anchorage, albeit in a modified out a study of the socio-economic dimensions of Yup'ik grass basketry.

During the first few years of the study, when Annie Don, my Yup'ik collaborator, and I traveled frequently throughout the Yukon-Kuskokwim Delta villages, we often arrived in communities with no prearranged place to stay. Almost everywhere, there would be an older person living alone who was willing to take us in.

Because it is unusual for Yup'ik / Inuit-especially older people-to live by themselves, I began to make inquiries. "Where's your daughter?" I'd ask. "Moved into Anchorage with the kids," would come the reply.

At the same time, I realized that Annie, who has lived in Anchorage since 1994, was part of this migration, as were her circle of mainly Yup'ik women friends. Furthermore, at the Alaska Federation of Natives Crafts Fair held in Anchorage each October, roughly 75% of the vendors are female.

So I wondered if the migration was evenly distributed between men and women. Sure enough, this demographic reflects the gender breakdown of Anchorage-based Alaska Natives as a whole. In 2000, there were slightly over 7,000 Alaska Native women 18 years and above residing in Anchorage, but only some 5,500 men.

How does urban in-migration affect Alaska Native women? For one thing, they confront on a daily basis political issues that would be at greater remove from them in the village. Some 50-75% of the urban-dwelling women support themselves through the sale of arts and crafts.

These commodities in turn depend for their appeal on exotic raw materials such as seal hide, walrus stomach, and basket grass only obtainable in rural areas. The question of whether urban Native people will have continued access to these products on public land is at present hotly debated and since their very livelihood depends on such access, they are beginning to participate more actively in the political process than would normally be the case. In 1998, the Alaska Federation of Natives, its collective patience worn thin by the absence of resolution, called for a public protest, summoning Alaska Natives to a march and rally in downtown Anchorage.

For Alaska Natives generally-and Alaska Native women in particular-direct confrontation such as a protest march, represented a lapse from culturally approved styles of conflict-resolution, which tend to favor settling disputes by negotiation. Among her circle of friends, the question of whether to participate was hotly debated. So as I circled by Annie's house to pick her up, I wondered if she would act.

But there she was on the doorstep. "It's for my grandkids," she commented as she climbed into the car, "Anyway if I can't go home and pick grass, how will I make my baskets?" Her remark illustrates the urban and rural connectedness that characterizes so much about life for Alaska Natives today. form emphasizing the common Yup'ik heritage rather than the local origin.

Thus, outmigration extends the social network of homeland villages and gives rise to "translocal communities"-at least for a time. A similar case has been made for the Inuit in Montreal, where it appears that urban Inuit are using ethnic identity as an adaptive tool in a multi-ethnic city.

Due to the original vocation of anthropology as a "western" discipline striving to study "exotic others," corresponding studies of southern immigrant communities in the Arctic are much less abundant. For instance, Danish identities in Greenland are only recently attracting the attention of researchers, one of them a Greenlander.

The persistence of kinship and community network among Arctic emigrants to urban areas is similar to recent observations in urban Southeast Asia, Latin America, and Africa. So far, the pattern is quite different from the developments that took place in Europe and in North America in connection with industrialization and urbanization.

Trend Summary

Most Arctic communities have experienced relentless and pervasive changes over the past decades. The disruption of communication between generations is as much a consequence as a cause of difficulties in social reproduction. Kinship ties and networks have not died out with urbanization but remain a focus for meaning and identity in the growing Arctic towns, and even in urban settings outside the Arctic homeland.

KEY CONCLUSIONS OF POLICY RELEVANCE

Arctic societies and cultures are highly adaptable and resilient and thus well-equipped for integrating change. The fact that they integrate modernity should be viewed positively rather than with nostalgia for traditions lost. The concept of traditions should be seen as a dynamic one: traditions do not and should not hinder development.

If there is a political lesson to be learned, it is that paternalistic attitudes, top down approaches, and change imposed from the outside are counter-productive, even if they seem to foster social and cultural perseverance in the short run.

In the long run, the transfer of decisionmaking powers from central to local authorities might be the most important factor in helping reduce social and cultural problems associated with rapid change. To reverse negative trends that have their origins decades or centuries ago, however, requires a lengthy "normalization" period. Thus, one cannot expect things to improve immediately.

In each Arctic region, there is a feeling of being distinct from regions further south, even where those regions are adjacent. Globalization increases worldwide connections, but the information and communication technology that fosters it is also used to intensify circumpolar connections.

The emergence of a pan-Arctic identity can be further encouraged. Finally, decision makers should not overestimate the predictive power of social science data.

The inherent complexity of social and cultural systems makes predictions that rest upon the extrapolation of particular factors or trends tentative at best. "Social planning" approaches of the past testify to these dangers. Only through further investments into the development of better datasets and methodological tool kits can these limitations be overcome.

Gaps in knowledge

Most Arctic states do not keep separate statistics for their Arctic regions regarding key indicators of social and cultural change. For example, it is virtually impossible to determine numbers of speakers of particular languages spoken in the Arctic or to find reliable data about the religious affiliation of Arctic residents. We therefore encourage the Arctic states and Permanent Participants to actively assist in the compilation and distribution of Arctic social science data.

The study of social and cultural processes in the Arctic has been dominated by the discipline of anthropology. While this has led to a number of excellent studies on the local level, regional and national data are much rarer.

The anthropological heritage of Arctic social science research has led to a particular emphasis on indigenous groups and peoples, at the expense of "mixed" and "newcomer" groups. Anthropological research in the former Soviet Union focused more strongly on processes of the past than in other Arctic countries.

Thus, there is a lack of studies regarding contemporary trends in the Russian North, especially for the period between World War II and 1990.

The former Soviet Union is not alone in putting more emphasis on certain kinds of research than on others. Most Arctic regions have excellent data on one or another aspect of social and cultural processes at the expense of others. A circumpolar research agenda would overcome the limitations of these "national" research agendas and result in increased comparability of the acquired data.

While there is ample documentation of the effects of individual aspects of social and cultural change, there is little understanding of the cumulative effects of rapid change. This gap in knowledge adds to the weak predictive potential of current knowledge.

summary

Arctic societies and cultures have been undergoing changes since time immemorial, but the 20th century resulted in an unprecedented amount and speed of social and cultural change. Still, even periods of rapid change have not eradicated traditional Arctic social and cultural systems. Rather, the contemporary Arctic is characterized by various combinations of "indigenous" and "western" elements. Alternatively perceived as "loss" or "innovation," these social and cultural processes inevitably lead to new forms of mixing "old" and "new." It will be up to Arctic residents to determine which kind of mix will best serve their future needs.

The Arctic has "grown" considerably during recent decades. Not only have contacts and levels of cooperation increased within the circumpolar North-especially since the opening of the former Soviet North-but the interconnectedness of Arctic and non-Arctic communities has become more apparent. While the impact of southern power centers on Arctic communities has long been noticed, Arctic communities are gradually expanding their reach south and thereby carrying Arctic social and cultural traditions into regions far removed from tundras and northern forests.

CIRCUMPOLAR SYNCHRONICITY DESPITESHORT-TERM OFFSETS

Within the overall trend of rapid social and cultural change, there is variation across the Arctic as to when these changes occurred. In Fennoscandia, indigenous peoples' contact with people of European origin started a millennium ago, whereas this happened only a century ago in remote areas of the Canadian Arctic. Developments in the Soviet and post-Soviet Russian Arctic have been quite distinct. Widescale changes started already a decade before World War II, with massive efforts to exploit nonrenewable resources, importing immigrant populations for that purpose. In addition, the state implemented ideologically driven experiments aimed at transforming society, especially with regard to indigenous peoples. It was not until 1990 that Russian indigenous peoples were allowed to create their own independent national organization: the Russian Association of Indigenous Peoples of the North (RAIPON).

Small indigenous peoples of the Russian North are now better connected with developments abroad than they were in Soviet times, but the evolution of their situation still remains quite distinct from other circumpolar regions.

Except in resource rich regions, this situation is characterized by dis-investment from the state, which previously played a central economic role. This has caused severe impoverishment and social and cultural disruptions.

In addition, there has been a massive out-migration of recent immigrants. Ever since the collapse of the Soviet Union, the Russian state has cut subsidies to its Arctic provinces, which has been very effective in inducing newcomers to leave as in Chukotka. Without such economic (dis-)incentives they tend to stay: Norilsk is a case in point.

Another significant area of contrast is that whereas most parts of the Arctic were Christianized before World War II, this process only began a decade ago in Northeastern Russia. During the 20th century, differential access to self-government introduced further divergences in social and cultural change across the Arctic.

Trend Summary

In spite of some differences in content and the timing of changes across the Arctic, some trends of the past few decades are amazingly similar and synchronous. Arctic cultures and societies are rapidly being transformed. After World War II, the rate of change accelerated and paternalistic welfare state policies were imposed. Such planned change gave rise to hitherto unknown social problems. Later, as the pressure to assimilate decreased, cultures and identities started to reaffirm themselves. This revival of tradition occurs in new forms that accommodate modernity and incipient globalization.

Bibliography

Alka Pareek: *Environment and Nutritional Disorders*, Aavishkar Pub, Delhi, 2003.

Anfinsen CB: *Advances in Protein Chemistry*. New York: Academic Press. 1972.

Banwari Lal: *Environmental Microbiology*, Cyber Tech Pub, Delhi, 2009.

Bhatia, S C : *Handbook of Environmental Microbiology*, Atlantic, Delhi, 2008.

Bose, Sarath Chandra: *Biochemistry : A Practical Manual*, Pharma Med Press, Delhi, 2010.

Bryant, Bunyan: *Environmental Justice: Issues, Policies, and Solutions*, Washington, DC, Island Press, 1995.

Chanderlekha Goswami: *Biochemistry and Instrumentation*, Manglam Pub, Delhi, 2011.

Cox MM: *Lehninger's Principles of Biochemistry*, New York, New York: W. H. Freeman and Company, 2005.

Daphne C. Elliott: *Biochemistry and Molecular Biology*, Oxford University Press, Delhi, 2005.

Dass Gupta: *Industrial Biochemistry*, Swastik, Delhi, 2007.

Florkin, M.: *Comprehensive Biochemistry: Amsterdam*, Elsevier, 1975.

Fumento, Michael: *Bioevolution: How Biotechnology is Changing Our World*, San Francisco, Encounter Books, 2003.

Garg, G.P. : *Biochemistry of Nucleic Acids*, Discovery Pub, Delhi, 2010.

Gay, Kathlyn: *Saving the Environment: Debating the Costs*, New York, Franklin Watts, 1996.

Goel, A K : *Basic Concept of Animal Chemistry*, Pearl Books, Delhi, 2008.

Goel, Gunjan : *Applied Dairy and Food Microbiology*, Agrotech, 2005.

Govindan, R. : *Environmental Microbiology and Instrumentation*, Manglam, Delhi, 2009.

Harsh Bhaskar: *Problems in Biochemistry*, Campus Books, Delhi, 2010.

Harvey L. Molotoch: *Urban Fortunes: The Political Economy of Place*, Berkeley, University of California Press, 1987.

Jack Simons: *An Introduction to Theoretical Chemistry*, Cambridge Univ Press, Delhi, 1998.

Jasra, O.P. : *Environmental Biochemistry*, Sarup & Sons, Delhi, 2002.

Kruvant, W.J.: *People, Energy, and Pollution*, Cambridge, Ballinger Publishing, 1975.

Mahaling, Ram Naresh: *Basics of Biochemistry*, Anmol, Delhi, 2008.

Mahboob, Syed : *Handbook of Fruit and Vegetable Products*, Agrobios, Delhi, 2008.

Meyer, Art: *Earth Keepers: Environmental Perspectives on Hunger, Poverty, and Injustice*, Scottsdale, Herald Press, 1991.

Ojha, Jai Shankar: *Aquaculture and Nutrition and Biochemistry*, Agrotech, Delhi, 2006.

Pathak, N : *Dictionary of Food and Nutrition*, Mohit Pub, Delhi, 2005.

Patricia Trueman: *Nutritional Biochemistry*, MJP Pub, Delhi, 2007.

Paul Mohai: *Race and the Incidence of Environmental Hazards: A Time for Discourse*, Boulder, Westview Press, 1992.

Prasad, Govind: *Environmental Geomorphology*, Discovery Publishing House, Delhi, 2008.

Prasad, S.K. : *Biochemistry of Proteins*, Discovery, Delhi, 2010.

Ruchi Gautam: *A Textbook of Microbiology*, Arise, Delhi, 2007.

Shubhrata R. Mishra: *Plant Biochemistry*, Discovery, Delhi, 2010.

Singh, Mahinder: *A Textbook of Biochemistry*, Dominant Pub, Delhi, 2011.

Suri, Nitin : *Molecular Biology and Biochemistry*, Oxford Book Company, Delhi, 2010.

Ullrich M *Bacterial Polysaccharides: Current Innovations and Future Trends* Caister Academic Press, 2009.

Wade, J. L.: *Society and Environment: The Coming Collision*, Boston, Allyn and Bacon, 1972.

Zupan, J.M.: *The Distribution of Air Quality in the New York Region*, Baltimore, Johns Hopkins University Press, 1973.

Index

❑❑❑